Insects of south-central Alaska

Dominique M. Collet

ISBN 978-1-59433-074-2
Library of Congress Control Number: 2008926170

Second Edition

Published by Kenai Watershed Forum

Design and production by Dominique Collet and Kathy Lepley, Galaxy Graphics
Printed in China

Table of Contents

Table of Contents

Table of Contents

Acknowledgements

This guide is the result of collaboration among many individuals. Ed Holsten initiated the project and remained involved through publication.

I am grateful to Pamela Houston for her support during the writing of the guide. Josselyn O'Connor and Robert Ruffner from the Kenai Watershed Forum managed the day-to-day details and provided necessary administrative assistance.

Guide development was primarily supported by the U.S. Forest Service. The U.S. Fish and Wildlife Service contributed to the aquatic section and the Kenai Spruce Bark Beetle mitigation program of the Kenai Peninsula Borough contributed the section on spruce bark beetles.

The originality of this guide is the product of the many scientists and amateur naturalists who shared their observations with me and allowed me to accompany them in the field.

Matt Bowser contributed to the identification of many insects photographed for this guide. Together, we always attempted to secure the identification of photographed insects by sending voucher specimens to specialized taxonomists. We received help with these identifications from Charles O'Brien, Charles Triplehorn, R. Gordon, K. G. A. Hamilton, Pat Bouchard, Ed Riley, John Luhman, J. P. Kopelke, Dan Bogan, Andy Graves, Brian Armitage, Steve Seybold, Christian Thompson, Robert Foottit, Terry Whitworth. I am grateful to the members of the steering committee and others whose comments and criticisms contributed to this work: Michael Fastabend, David Wartinbee, Mike Gracz, Lois Bettini, John Delapp, Ed Berg, Ken Zogas, James Kruse, John Luhman, Derek Sikes, John Morton, Roger Burnside, Judy Jacobs, Andrea Medeiros, Corlene Rose, Janice Chumley, Cathy Turner, and Dave Allen.

Some photographed specimens were provided by Crystal Collet, Rob Lipkin and the late Carol Lloyd.

Barb Maynard reviewed the manuscript. Kathy Lepley assisted with the design of the book and completed the layout.

I assume responsibility for any errors.

2009 Dominique M. Collet

Slopes of Manitoba Mountain on the Kenai Peninsula. The alder aphid can reach high densities in the extensive alder thickets covering the mountain slopes.

Foreword

I first became aware of Dominique many years ago after obtaining a copy of E. Viereck's 1986 *Alaska's Wilderness Medicines.* It was quite apparent that Dominique was a phenomenal artist. Years later, I received a call from Dominique. He wanted to come to Anchorage and look over our forest insect collection. He was investigating the parasitoids of willow gall makers as well as identifying a very complex insect family, the leaf beetles. After a brief discussion I realized that not only was Dominique a phenomenal water-colorist of the natural world, but also a naturalist and entomologist, par excellence! Dominique's latest products, guides to the willows of south-central and interior Alaska, beautifully demonstrate Dominique's blend of artistry and understanding of Alaska's flora and fauna. Further discussions and interactions with Dominique led to a discussion on producing a book on the common insects of south-central Alaska.

Why the need for a book on common insects of south-central Alaska? First, insects are the most common and diverse organisms on the planet and have been around for hundreds of millions of years. They occupy almost every possible ecological niche, from the tropics to the arctic regions. Most insects are beneficial to man providing pollination, recycling of plant and animal matter, and providing beauty in motion. Who has not been enthralled by the flight of beautifully colored butterflies or dazzling dragonflies? Alaska is a land of magnificent landscapes, plants, and animals. We are commonly reminded in everyday life about insects, from spreading honey on our toast in the morning to swatting mosquitoes and black flies throughout our long Alaska summers. One can venture into any local bookstore, museum, or visitor center and browse through dozens of books covering large mammals, birds, fish, wild flowers, mushrooms, and edible plants. There are, however, no books on the common insects of Alaska, let alone those of south-central Alaska.

For readers who, like myself, have been interested and enthralled with insects since childhood, this book will be a joy to own and use. For others, whose curiosity about some unknown creature encountered on a hike, while fishing or perhaps just in the backyard, drives them to seek information, this field guide will allow them to become acquainted with the marvelous world of "bugs." This long overdue book allows all of us, even the seasoned naturalist, to look at and begin to understand insects as wonderful co-inhabitants of our world.

Dr. Ed Holsten
Research Entomologist, retired
Cooper Landing, Alaska
August 2005

Introduction

How many species of insects are there in south-central Alaska? No one knows. Only a few groups of insects have been studied well enough to estimate the number of genera or species in the area covered by this guide: bark beetles, day butterflies, ground beetles, mosquitoes, and a few more. Most other groups have been only partially surveyed, with many species still to be recorded. Approximately 1200 species of beetles, 158 species of true bugs, 30 species of dragonflies, 80 species of butterflies, etc. have been identified for the entire state of Alaska. Attempts to develop a comprehensive inventory have just begun, but the total number of insect species in Alaska may well exceed the 5400 species estimated for Canada's Yukon Territory.

Most in-depth entomological studies in south-central Alaska are likely to bring attention to species unknown to the state, Canada, North America or even species new to science. Dr. David Wartinbee, Professor at Kenai Peninsula College in Soldotna, studied the empty pupal skins cast off by emerging adult midges (Chironomidae) found floating in the main stem of the Kenai River. Wartinbee identified 72 species of midges new to the region. His study increased the number of known midge species for south-central Alaska from 14 to 86. Of these, six or more species are new to science.

Even for most of the species that have been named, little is known about their biology. So much still remains to be discovered! Dedicated professional and amateur entomologists have many opportunities to contribute to the knowledge of insects in Alaska.

How To Use This Book

This book is organized into two sections. The first section, **General Insect Biology,** covers general facts about insects and some important aspects of their biology. The second section, **Insect Classification and Identification,** introduces the reader to common or remarkable species found in south-central Alaska.

Under **General Insect Biology,** facets of insect biology are first presented, then insects are grouped according to their life strategies into guilds. Next, the book looks at the association of insects into communities. Finally, some examples of outbreaks are discussed. Under **Insect Classification and Identification,** insects are organized according to their orders, families and genera. An attempt was made to cover insects that are most likely to be encountered. A key to orders is found at the beginning of the Part 2. The natural size of the adult insect is shown either as a silhouette inside a circle or the magnification of the main illustration is shown.

Collection Techniques

Various techniques are used to capture insects, from simply plucking insects from a window pane to using specialized equipment.

Butterflies and other flying insects are caught with an **aerial net.** This net is made of mosquito netting 2 feet long fitted onto a stiff circular frame 10 or 12" in diameter attached to a pole. The bottom of the net should be rounded and not extend in a long point. The **sweep net** is a tougher version of the aerial net and is used by sweeping through the tops of shrubs and grasses in a figure 8 pattern, keeping the net opening facing the forward direction. This net can be homemade with soft 1/2" copper pipe for the frame and handle and bolting cloth for the bag.

Terrestrial insects such as ground beetles are found under large stones, logs or flat objects lying on the ground. An empty plastic sour cream container buried with the top rim flush with the soil surface makes a fine pitfall trap. One half inch of water with a squirt of dish soap is poured into the container to trap and kill the insects. These traps should be checked regularly and the insects removed before they start to decay. A small piece of plywood or a paper plate mounted above the container will keep it out of the weather.

An empty coffee can, with small holes punched through the bottom for drainage, buried with the open top flush with the soil surface, also works well. This trap can be baited with a piece of raw chicken. When "ripe," the bait attracts colorful red and black carrion beetles.

In south-central Alaska, the short summer nights limit the efficiency of **black light traps.** Fluorescent black lights attract many moths and aquatic insects that disperse at night. Black light traps sold in specialized entomologist supply outlets come with a light sensor and are powered by a small car battery. But it is not necessary to have such a complicated contraption. A light in front of a spread white sheet will attract many insects that can be plucked off the sheet. Ceiling light fixture covers are also worth checking for dead insects. All year long the lights attract flying insects, which are usually very well preserved, though often dusty. Out-of-reach windows sills are excellent places to find insects in the dead of winter.

Aerial net

The **malaise trap** is a single-pole tent made entirely of mosquito netting. The tent is partitioned by two vertical intersecting panels forming four inside walls. Openings at the peak of the walls, under the pointed roof, funnel the intercepted insects into a canister perched inverted at the top of the pole. A strip of insecticide kills the insects collected in the canister. These passive traps intercept actively traveling insects such as generalist predators and parasitoids in search of their quarry, pollinators, adult insects in search of mates and mated females in search of sites to lay their eggs.

Other passive collecting devices can be tailored for specific insects. For instance, adult spruce bark beetles and associated insects are captured with **Lindgren funnel traps:** these consist of a set of a four to sixteen black funnels linked together, hanging vertically a few inches apart. The whole contraption, 4-5 feet in length, hangs from the limb of a tree. A small bait vial attached to one of the funnels slowly dispenses a synthetic imitation of the female spruce bark beetle's aggregation pheromone. Spruce bark beetles and other associated insects are attracted to the scent and try to land on the trap. They land on one of the funnels, lose their grip and fall from one funnel to the next through the central opening, ending up in a collecting jar at the bottom.

Berlese funnels are used to capture the minute insects that live in the soil. A sample of duff is left on a screen above a broad funnel placed on top of a collecting jar. A heat lamp placed above the soil slowly dries the surface, thus making it inhospitable for the small insects which are forced to retreat into the darker, cooler underside of the soil. They then fall into the collecting jar.

Other collecting methods include spreading a large sheet of white cloth under a tree or a shrub and shaking the foliage above to dislodge the insects, which can be picked up on the sheet.

Sedentary species or larval stages are not sampled well by these passive traps.

Malaise trap

Black light trap

Spruce bark beetles are found outside their galleries for only a few weeks each year, just long enough to mate and disperse. For the remainder of the year, they hide under the bark of spruce trees. The various life stages can be collected by carefully lifting sections of the bark of affected trees.

Lindgren funnel trap

Preservation Techniques

Captured insects are killed by freezing, by leaving them in a killing jar or by dunking them in alcohol. Deep-freezing overnight will kill most insects captured during the warm season. This method of course does not work for those insects that are ready to overwinter or for a number of arctic or alpine species.

If acetone (nail polish remover) is used to kill the insects, it is either soaked into 1/2" of plaster of Paris previously poured in the bottom of the jar or it is soaked into a paper towel kept in the killing jar. Caution should be exercised as these chemicals are also toxic to humans. Please read and follow instructions on all labels. The insects should be left in the jar with sufficient chemical long enough to be killed.

Swept material can be dropped into a jar containing alcohol and sorted out later, under magnification. The alcohol will help preserve most groups of insects.

Small insects should be handled with tweezers or better yet, with the tip of a moistened watercolor paintbrush.

Many of the insects encountered in south-central Alaska are small, with most less than 1/4" long. A good dissecting microscope is necessary for any serious study. A binocular dissecting microscope magnifies objects 5 to 40 times and gives a three dimensional view of small objects. The object does not need special preparation. Compound microscopes, in contrast, magnify an object 40 to 1000 times, but require the object to be transparent and prepared on a microscopic glass slide for viewing.

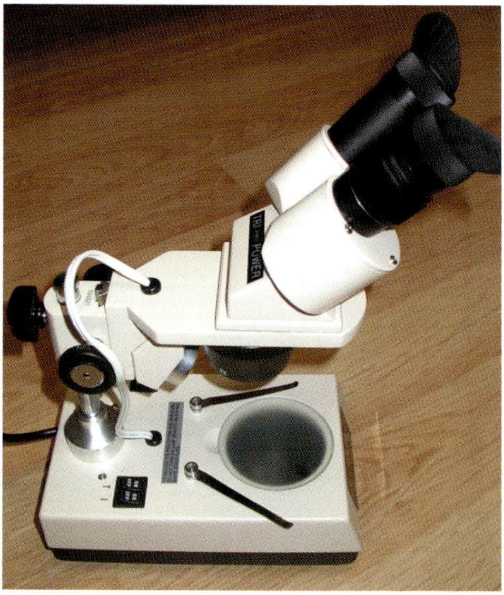
Dissecting microscope

The best way to preserve insects is to pin them. Insect wings and legs can be spread at the time of pinning and then the insect must be left to dry for several weeks. Once pinned, the insect can be handled by the head of the pin. Dried insects are brittle and easily damaged if touched. Minute insects are glued onto the tip of a paper point

Pinned insect

Paper pointed insect

mounted on a pin. The insect is glued on its right side, with a minimum of glue so as not to coat the leg articulations.

Pinned insects are kept in a flat box lined with a 1/2" thick layer of soft material. A sheet of styrofoam or flat ceiling tile cut to fit the bottom of a cigar box or a metal cookie box will do the trick.

At the time of capture a brief note is written in black pencil (not ballpoint pen) answering the questions "who, where, when and how" (who captured...). This information is copied onto a label, which is then attached to the specimen. Insect specimens with reliable information regarding the collection can be valuable for research even if they are not identified. **It is important to include the latitude and longitude coordinates of the location where the specimen was captured.** Additional information on the biology is very useful: name of the plant or the animal the insect was found on, description of the habitat, latitude and longitude coordinates, behavior observed, etc. The insect may also be given an index number referring to observations kept in a separate notebook.

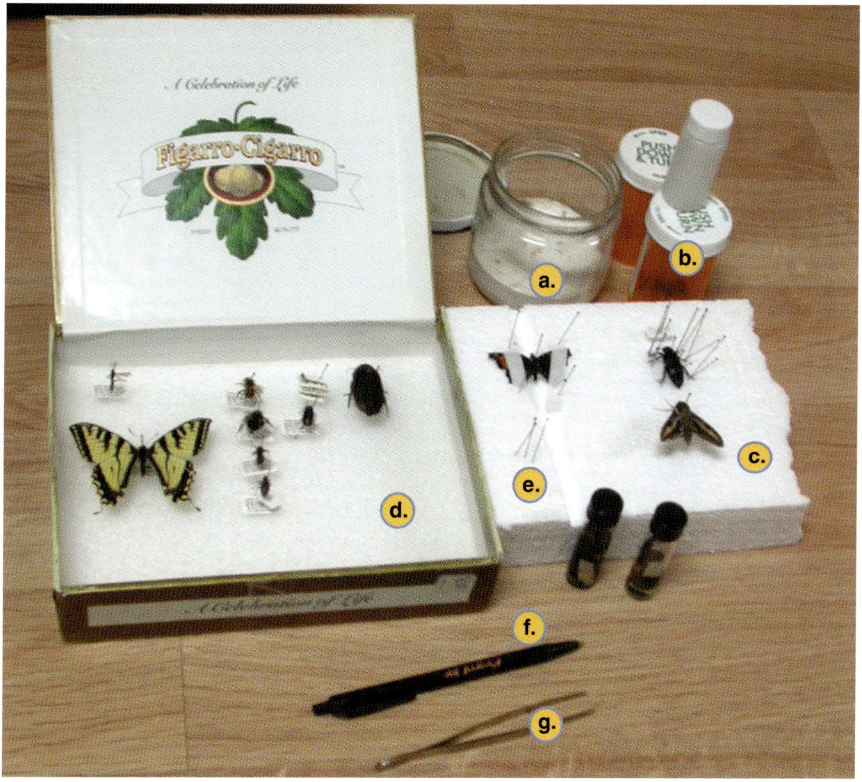

Basic collection kit

a: killing jar: 1/2" of plaster covers the bottom of a screw-cap glass jar. The dried plaster of Paris is impregnated with acetone (nail polish remover), **b:** collecting jars, **c:** spreading board: piece of styrofoam from packaging material with a groove cut in the foam to receive the body of butterfly while spreading the wings, **d:** cigar box lined with a 1/2" thick sheet of soft material such as styrofoam or acoustical tile covered with a sheet of white paper at the bottom, **e:** pins, **f:** lead pencil and **g:** fine tweezers.

Physical Setting

South-central Alaska encompasses a large region that lies south of the Alaska Range, including the Matanuska and Susitna Valleys, Kenai Peninsula, the western side of Cook Inlet and the coastal area of Prince William Sound.

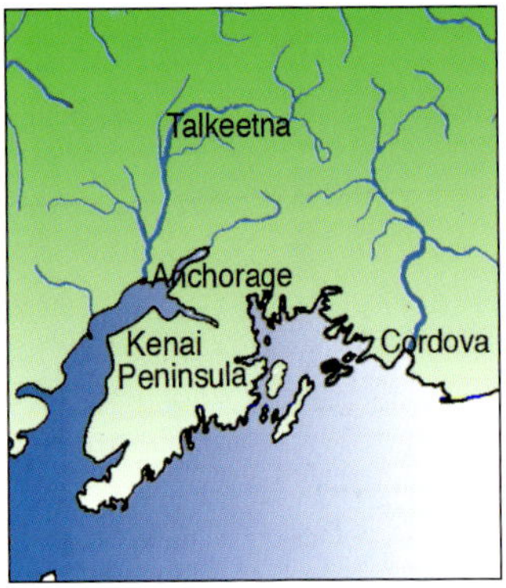

Area covered by the guide

The terrain is physically diverse. Landforms include broad valley bottoms, mountain slopes and elevated plateaus, rocky outcrops towering above alpine ice fields, extensive mudflats exposed at low tide on the shorelines of the upper Cook Inlet, and the wave-beaten rocky shores of Prince William Sound.

The climate of the coastal regions is driven by the mass of the North Pacific and is characterized by high precipitation, cool summers and mild wet winters. The regions that lie in the rain shadow of mountain ranges experience lighter precipitation and greater temperature contrast between hot summers and cold winters.

Long-term Climatic Effects

This land is new. Twenty thousand year ago, a blink of an eye in geological time, most of the region was encased in ice, with limited sections of land such as the Caribou Hills north of Homer remaining free of glaciation. Most, if not all, terrestrial life was extinguished in the glaciated area or forced out during the ice age.

The closest ice-free land from which terrestrial species recolonized at the end of the most recent glaciation lies north of the Alaska Range. The transitional forests of the south-central Alaska lowlands are connected to the boreal forest, which covers the rest of Alaska, through two broad mountain passes in the Alaska Range. The Parks and Glenn highways follow these routes through the passes at elevations nearing the limit for tree growth.

The Kenai Peninsula, Prince William Sound coastline, and the Kodiak archipelago harbor the northwestern reaches of the coastal temperate rain forest, reduced to a few cold-hardened species. To the south, the coastal rain forest of northern California is the most diverse with fourteen species of conifers, of which only four make it to south-central Alaska. This young land is in a state of flux. As global warming sets in, one can expect the range of coastal species to extend northward.

Vegetation

The flora of the south-central Alaska lowlands combines elements of two major North American ecosystems: the coastal rain forest and the northern boreal forest.

Sitka spruce, western hemlock and balsam poplar of the Prince William Sound coast define the coastal rain forest while white spruce, black spruce, paper birch, balsam poplar and aspen found in Western Kenai Peninsula lowland and the Matanuska Valley are the typical elements of the boreal forest. Frequent fires and spruce bark beetle infestations rejuvenate the mature vegetation, allowing new growth to develop.

Biodiversity

Cold temperatures forbid the establishment of many temperate species. The extensive boreal forest is made up of a handful of tree species. In northern regions, many unique floral and faunal assemblages have been extinguished by the repeated glaciations. Expanding glaciers scoured valleys during each of these glaciations, and at the end of each glacial event, the receding ice dumped the burden on the lowlands, reducing the region to a barren moonscape. The contemporary flora and accompanying fauna for whole regions descend from a few surviving species representing a fraction of the previous biodiversity. These few survivors followed the retreating ice and flooded the new landscape. Contemporary boreal flora and fauna are impoverished compared to tropical forests whose rich biodiversity results from long, uninterrupted benevolent climatic conditions that allowed for the formation of complex forest structure over time. In tropical settings, speciation occurs under evolutionary pressure, while extinction remains low because of the stability of the climate.

A mere trickle fed by snowmelt up where willows thickets dominate, the Russian River is all whitewater as it roars down the Falls. Adult stoneflies emerge in great numbers from these pristine waters of the Kenai mountains.

The author searching for bristletails near the headwaters of Eagle River, in Chugach State Park. Bristletails thrive among the lichens colonizing the fractured boulders jammed at the base of the steep rock walls blocking the head of the valley.

Subalpine meadows with scattered willows in the Kenai Mountains. The white phoebus parnassian, a true alpine species, ventures down this vegetation zone. Summit Lake is in the background.

These coastal sand dunes near the mouth of the Kenai River are stabilized by grasses and shrubs. Willow leaf beetles can be abundant on willows and cottonwoods.

The four-spotted skimmer, Alaska's state insect, is found in habitat such as this roadside pond near Tern Lake on the Kenai Peninsula. The predatory larvae hunt among submerged vegetation.

Part 1

General Insect Biology

Body Parts

The body of adult insects is divided into a head, thorax, and abdomen. Three pairs of legs are attached to the middle section, the thorax. The head bears the antennae, mouthparts and eyes. The abdomen contains the digestive and the reproductive organs. Around this basic structure, insect anatomy varies greatly. Most adult insects have two pairs of wings attached to the thorax. Wasps (Hymenoptera) have large forewings and smaller hindwings, while beetles' (Coleoptera) have forewings that are hardened into elytra protecting the functional hindwings, and flies (Diptera) have only one pair of forewings and the hindwings are reduced to short stubs (halteres). used during the flight.

Hymenoptera

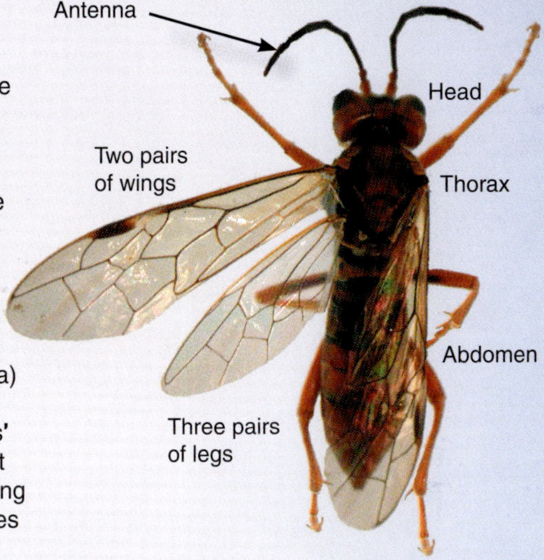

Sawfly Wasp
***Tenthredo* sp.**

Coleoptera

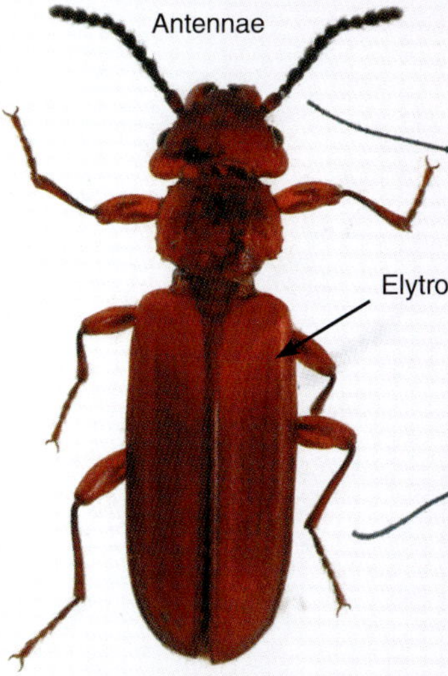

Cucujid Beetle
Cucujus clavipes

Diptera

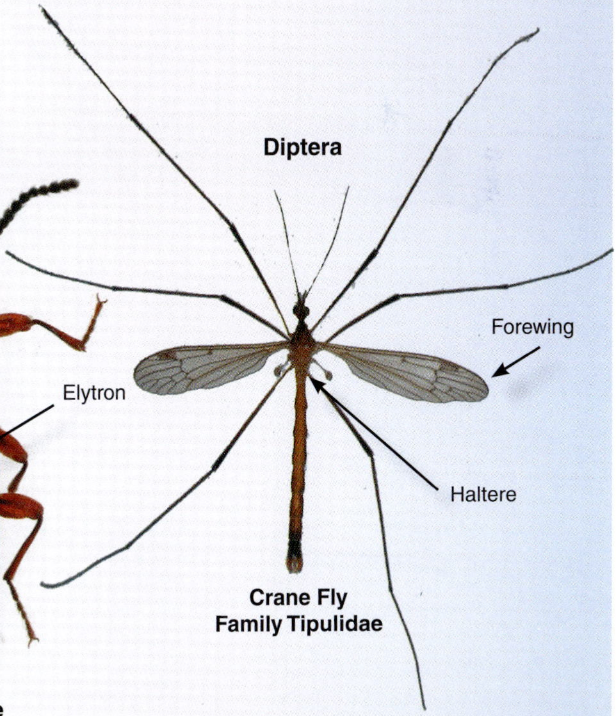

Crane Fly
Family Tipulidae

Examples of Insect Mouthparts and Antennae

Insect body parts vary widely in shape and size. The mouthparts can be highly modified to fit a certain function, or in some cases, can be missing entirely, because the adults do not feed. The antennae can vary from a simple wiry shape to feather-like.

Immature insects grow in increments. They live through a series of 3 or more molts during which they shed their rigid chitin carapace. Chitin is the substance of our hair and nails. A new carapace is formed by the exposed "plastic" under-layer which stretches to accommodate the increased volume of the body before hardening.

Immature insects follow one of two courses of development: they either go through incomplete or complete metamorphosis.

Incomplete Metamorphosis

The immature stages, called nymphs, are similar to the adult. For species whose adults have wings, wing pads are present from immature stages and keep developing externally with each molt. There is no transitional pupal stage between the nymphs and the adult insects. Nymphs often have life habits similar to the adults although many immature aquatic insects with incomplete metamorphosis are aquatic while the adults are aerial.

Dragonflies, aphids, mayflies, grasshoppers and true bugs go through incomplete metamorphosis during their development.

Birch Shield Bug
Elasmostethus interstinctus
The shield bug nymphs resemble a diminutive adult.

Complete Metamorphosis

The larvae are very different from the adult and the wings develop only at the adult stage. Major reorganization of larval tissues into adult organs (metamorphosis) occurs during a special resting stage, called pupation.

Beetles, flies, wasps and butterflies go through this type of development.

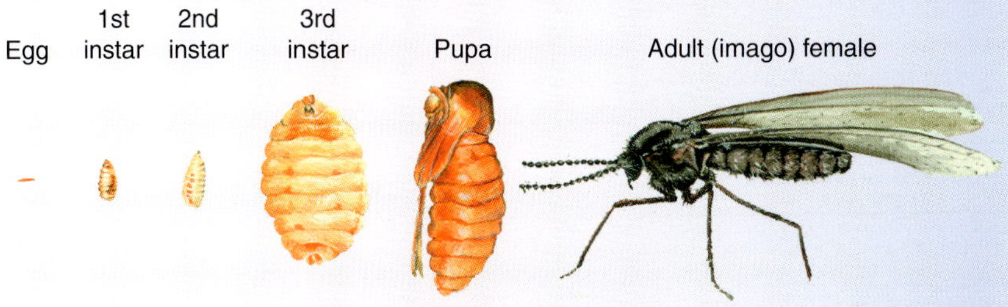

Willow Rosette Gall Inducer
Rabdophaga rosaria
The life cycle of gall midges is typical of insects with a complete metamorphosis. During its development, the insect passes through four distinct phases: egg, larva, pupa and adult.

Insects basically have the same senses as humans. They see, hear, taste and smell, but some of their functions differ. For example, for the smaller insect species, the sense of sight is minimized while the sense of smell is significantly increased in importance. The quality of the image transmitted through the eyes diminishes with their size; the function of miniaturized eye organs called ocelli is reduced to that of a light sensor, with no detailed image perception. On the other hand, the sense of smell is of paramount importance because odors carry meaningful information over long distances.

Vision

Immature insects have simple eyes called ocelli[1]. The number of pairs of ocelli varies among insect groups. Caterpillars have six ocelli arranged in a semi-circle on either side of the head, while sawfly larvae, which are similar in general appearance to caterpillars, have only a single ocellus on either side of the head. Ocelli are crude light sensors that detect changes in light pattern. Adult insects often have three ocelli arranged in a triangle at the top of the head between the two compound eyes, which are the only functional eyes. For insects that mostly live underground, and even for a few that live in the open, vision is not essential. Their eyes are either underdeveloped or are shielded under some extension of the carapace.

The compound eyes can be large. Their outer layer is composed of closely packed truncated cones (ommatidia). Each cone has a lens at the outer, wide end and a light sensor at the inner, narrower end. The lenses appear as a honeycomb pattern on the eye surface. Each lens focuses the light onto one sensor. The information gathered by each of these sensors can be compared to that of a single pixel on a computer screen. The composite information from all cone sensors constitutes the vision of the insect.

The number of facets of a compound eye varies from as low as 6 for some ant workers to as many as 25,000 for a dragonfly which corresponds to a resolution similar to an image of 158 by 158 pixels, or about 2"x 2"on a computer screen monitor. Compound eyes are very efficient at tracking movements, but the keenest insect vision still has only 1/60th of human acuity.

The cones are not evenly distributed on the eyeball. They are denser in sectors according to the insect's life history. The higher the number of

1 Ocellus: singular, ocelli: plural.

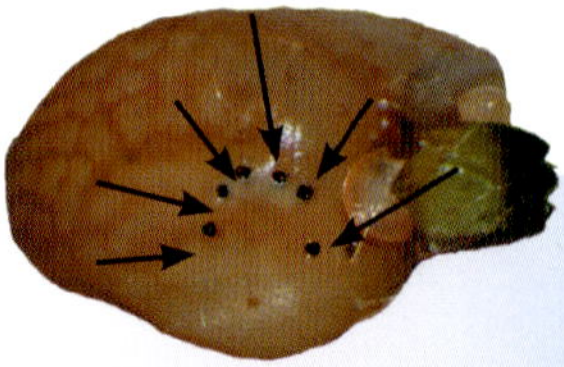

Side view of a caterpillar head

Caterpillars have 6 ocelli in a half circle on either side of the head.

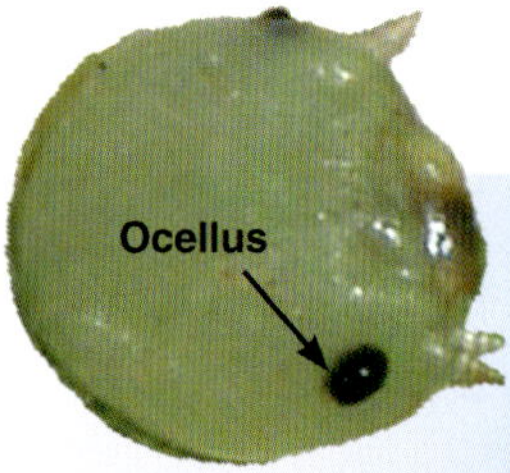

Sawfly larva head

The sawfly larva's small ocellus is centered on a flat, often dark-colored disk that makes the ocellus appear larger than it actually is.

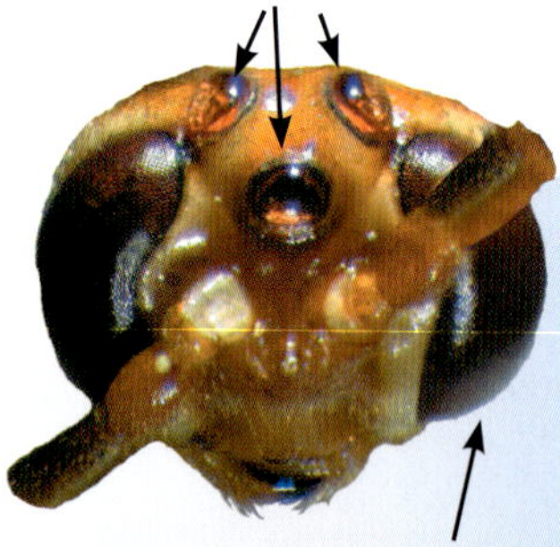

Adult ichneumonid head
The three ocelli are arranged in a triangle at the top of the head.

sensors facing a given direction, the higher the definition the perceived "picture" will be. For example, diurnal predatory insects such as dragonflies have more sensors in the forward-facing part of their eyes.

The upper half of the eyes of male mayflies and danceflies are extremely developed, covering the upper half of the head, almost splitting away from the lower lobe of the eye, which is by itself equal to the female's eye. This "hypertrophy" is correlated with mating behavior for these two groups of flying insects. In both of these insect groups, adult males congregate in dense cloud-like swarms. Females are attracted to this display, approaching the turbulent cloud of males from above. With their "super" upper eyes lobes, males can discern the females in flight as dark silhouettes against the sky. As soon as the females come within visual range, males rush to meet them. Judging from the exaggerated development of the upper half of the eye, males with the keenest vision must have had a competitive advantage that translated into the fixation of genes that carry the odd hypertrophic mutation of the male's eyes.

In the case of big-headed flies, it is the females that have extraordinarily developed eyes. The compound eyes cover the surface of the entire head! This can be correlated to the need for the parasitic big-headed fly females to locate their leafhopper hosts while hovering among grasses.

Whirligig beetles that spend their adult life spinning on the surface of standing water have split vision. The upper half of the eyes allows them to see above water while the lower half is corrected to see underwater (see p. 65).

Insects use their vision for navigation at night. By keeping the moon aligned with a few cones of their compound eyes, they keep their head in a steady orientation and travel in a straight line. In the natural world, using the moon works well because it is a distant beacon. But if an artificial light source such as a candle or an incandescent light is substituted for the moon, the same insects will not travel in a straight line but follow a spiral trajectory that will lead them to the light source.

Dance Flies
***Bibio* sp.**

The upper side of eyes of the males are extremely developed for quickly spotting females against the sky

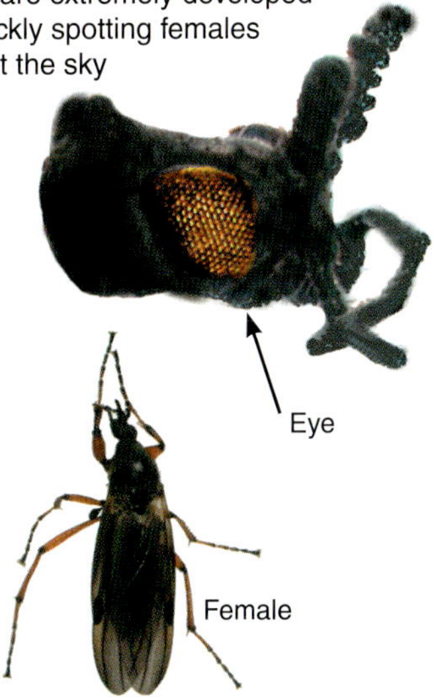

Big-headed Fly
Pipunculidae

Smell

The sense of smell, or the ability to detect and identify volatile chemicals, is very developed for most insects. Airborne chemicals are constantly dispensed into the environment. The most trivial biological activities are accompanied by the release of some form of chemical signature, information that potentially can be gleaned by interested parasitoids and predators.

The way odor disperses in the environment can be visualized as a plume of smoke: concentrated and narrow at the source, diluted and spreading as it drifts away. Insects entering the plume of an attractive volatile chemical locate its source by zigzagging across the plume, sensing the ever so slight increase of the compound in the air.

Just as we can filter out a conversation from a noisy crowd, insects can detect important odors in the cacophony of chemical compounds that flood their environment.

Larval parasitoids of the white cabbage butterfly locate caterpillars by detecting airborne chemicals released by leaves whose cells are ruptured by the feeding caterpillar, combined with a signature chemical of the saliva of the caterpillar.

Female moths give off pheromones[1] to attract males for mating. The organ of smell is located in the antennae. Male moths and mosquitoes have feather-like antennae whose extra surface area is covered with odor sensors that increase the sensitivity to pheromones, allowing them to detect minute amounts.

As soon as it emerges from its pupal stage and before it exits the cocoon, the flightless adult female tussock moth releases pheromones, luring males. The male tussock moth has normally developed wings, giving him a very different appearance than the female. His antennae, where odor sensors are located, are branched to the point of resembling a feather.

Parasitic Wasp
Dendrocerus carpenteri

The importance of the sense of smell for many small insects is underscored by the disproportionate size of the antennae relative to the body size.

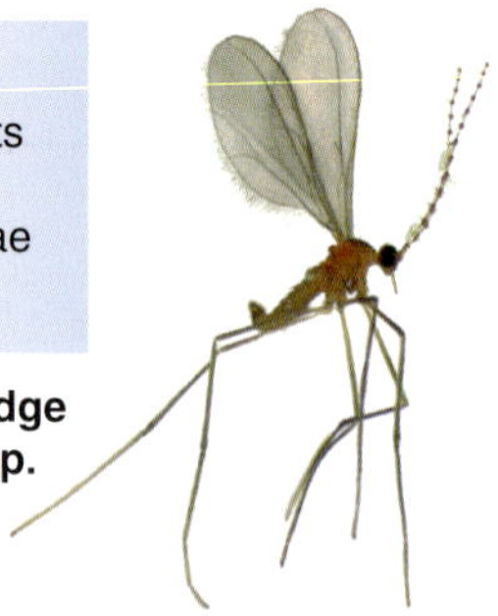

Small Gall Midge
***Dasineura* sp.**

1 Pheromones are chemicals emitted by an animal to communicate with other members of the same species.

Sound

Few insects in south-central Alaska emit remarkable sounds. Perhaps the noisiest are feral domestic crickets, survivalist escapees from pet stores that end up hiding in public buildings, serenading all night from behind sheet-rock walls or from within heating ducts.

Their wild relatives, the grasshoppers, produce only modest stridulations.

Each mosquito species buzzes at a specific frequency recognized by potential mates.

The feather-like antennae of some male no-see-ums act as auditory organs and are sensitive to the wing beat frequency of an incoming female. Possibly, the natural frequency of the antenna or parts of the antenna matches the wing beat frequency of their potential mate. If so, the auditory organ of the male would vibrate intensely as a female approaches.

"Deathwatch beetles" are beetles that bore into wooden furniture. At mating time, they bump their heads against the gallery wall. In a quiet room, the sound is audible as a ticking.

Cricket
Acheta domesticus

Occasionally, in south-central Alaska, one can hear feral crickets in heated buildings.

Stegobium paniceum

The drugstore beetle, *Stegobium paniceum*, is closely related to the deathwatch beetle. This species is a pest in stored goods.

Touch

When searching for their hidden hosts, parasitoids often touch or tap with their antennae the surface of a plant where their host might be hiding. Perhaps they are able to locate their host by sensing vibrations.

The body of many insects is covered with stiff silks and hairs, sensors that inform the insects about their surroundings. The long forward-oriented antennae of the cricket and other nocturnal insects probe the terrain for obstacles.

Coloration

Camouflage

Some insects blend in with the background while others actively stand out.

The *Syngrapha* moth is a champion in the art of camouflage. When resting on birch bark, it becomes just another rough scar.
Freshly emerged moths even have tufts of hair on the back that further break the outline of the body.

Flowerflies pupate in the foliage of the tree where they were feeding on aphids. Their pupae remain glued to a leaf and resemble bird droppings.

Aposematic Colors: WARNING

Insects that secrete toxic chemicals or are armed with an appendage that delivers a painful sting gain advantage by standing out and being obvious. Predators that have had a prior unpleasant encounter will likely avoid repeating the experience, if given a chance.

As a general rule, insects that are easy to spot against their surroundings are protected from generalist predators: the brightly contrasting yellow and black pattern on the abdomen of paper wasps warns all against considering the wasp as a likely meal. The orange-colored elytra of ladybugs warns would-be predators that the beetle regurgitates a nasty liquid when attacked. Yellow-and-black stripes and bright orange patterns are aposematic: they are universally understood as warnings for danger both in the natural and in the human world, as testified by the use of warning tape and barricades posted by police or highway maintenance crews.

Yellowjacket
Vespula vulgaris

Lady Beetle
Coccinela transversoguttata

Mimicry

A few insects gain some protection from predators by mimicking other insects that have a bad reputation. For example, flower flies resemble yellow jackets or bumble bees, clearwing moths mimic wasps, and botflies look like bees. In these cases, all the models are Hymenoptera and can be distinguished from their mimics by the two pairs of transparent wings, and the elbowed antennae. Flies have only one pair of wings and very short antennae. Clearwing moths have long club-shaped antennae and sections of the wings covered with minute scales as do all the other moths.

Insect Survival

In a stable insect population, for each actively reproducing pair, only two of the descendants, on average, will survive to reproductive age regardless of how many offspring are generated.

The mortality rate that can be expected for each generation to procreation age can be easily estimated: it is the number of offspring generated by one breeding pair minus two. Thus, to know the average number of progeny produced during the lifetime of a female of a given species is to know the mortality rate expected for that species.

On average, insect females will lay 100 to 150 eggs during their lifespan.

In contrast, female louseflies give birth to a total of 7 or 8 larvae ready to pupate. In this case, the remarkably low reproductive rate can be explained by the fact that the fly skips altogether the autonomous larval stage, thus greatly reducing the mortality, and spends most of its adult life sheltered within the plumage of its bird host.

We could also contrast this cozy lifestyle with that of aphids. These plump critters are thin-skinned and sluggish, live exposed on leaves and when feeding with their stylet deeply imbedded in plant tissue are unable to escape. Their reproduction rate is phenomenal. For part of their life cycle, aphids skip sexual reproduction. Without mating, females (fundatrix) produce on average 1.6 females per day, each one also capable of reproducing at the same rate at the age of 7 days old! At this rate, if there was no mortality, a single fundatrix could generate more than 2,000 aphids in 21 days.

Potential growth of an aphid colony generated by a single fundatrix

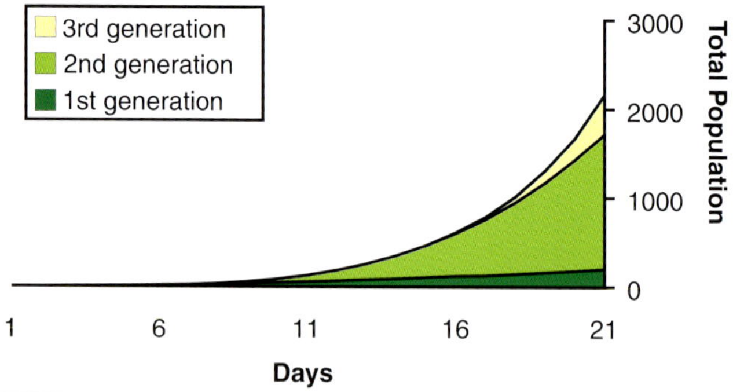

This can be observed in a natural setting where an aphid colony will at first double about every other day until predators and parasitoids catch up and feast on the easy prey. Predators soon decimate the population, leaving a few stragglers by the end of summer.

Cold climates present a challenge for the survival of living organisms in two distinct ways: the reduced amount of solar energy available during the growing season and the lethal effect of the lowest temperature reached during the winter. These two aspects of cold climates affect living organisms independently.

The growing season, defined here as the period between the last frost in spring and the first frost in the fall, lasts 4 to 5 months in the Kenai Peninsula lowlands. The low solar intensity is somewhat compensated by the long daylight hours. This growing season shortens with a rise in elevation. Along the coast, the cold oceanic mass generates a cool inland breeze and increases the cloudiness that depresses the average daily summer temperature even further. The short duration of the growing season coupled with the low average daily temperature result in a low summer solar gain. Compared to temperate regions, local insects have an abbreviated life cycle, fewer annual generations, or have generations spanning two or more years instead of a single season. Particularly affected are those species whose life cycle is synchronized with the life of other organisms.

Temperatures may drop lower than -40° F in depressions on the Kenai Peninsula lowlands and in the Alaskan Interior. At these low temperatures, water crystallizes in almost all living tissue of insects. Three major strategies are utilized by cold climate insects to survive the extreme low air temperature: avoidance, resistance to tissue freezing, and tolerance to freezing solid.

Frost is lethal for most insects of temperate regions. However, these insects can survive in south-central Alaska if they manage to find frost-free pockets such as lakes and flowing rivers. The ground under the river bed of running creeks never freezes even at locations where the temperature drops well below freezing for extended periods of time. Snow also has excellent heat insulating properties. Insects overwintering in the shallow duff layer rely on an undisturbed snow cover for the duration of winter to prevent deep frost from reaching them. For many species, winter snow conditions are the most important factor in their population dynamics. Insects that overwinter above the snow pack, such as bark beetle larvae under thin bark or wasp larvae in galls persisting on shrubs, are exposed to the coldest temperatures. Either their body fl uids never crystallize or they are freeze tolerant. In the first case, some larvae have developed a remarkable ability to resist freezing by converting some of their body fluids into antifreeze. The larva of the beaked-gall inducer *Rabdophaga rigidae,* for example, does not freeze when exposed to temperatures as low as -59° F. In winter, 10 to 12% of its body mass is glycerol, a natural antifreeze.

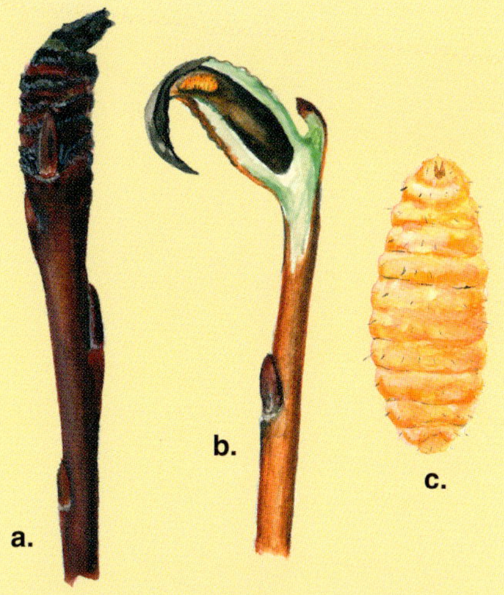

Rabdophaga rigidae. a. gall, b. cross section of the gall containing the overwintering larva, c. overwintering larva. Experiments have shown that *R. rigidae* larvae can survive exposure to -59° F.

Other insects that are exposed to low temperatures revive after having been frozen solid.

The body temperature of insects changes with the temperature of their surroundings. A few cold climate insects have a metabolism that allows them to remain active at temperatures lower than that of insects from temperate zones, an adaptation that extends their active season.

Bumble bees, for example, generate their own body heat by vibrating their wings before taking flight. Their body is covered with an insulating layer of hair similar in structure to bird down, with each hair divided into multiple branches.

Insects overwinter in different life stages: aphids overwinter as eggs, tiger moths and willow rosette gall midges as larvae, swallowtails and mustard whites as pupae, and thrips, willow leaf beetles, mourning cloaks, bumble bees, and willow weevils as adults. Most go into a coma-like state called dormancy, during which all the metabolic functions are reduced to very low rates. Dormant insects do not become active as soon as they are warmed up, but need some time for their normal functions to resume.

Gardeners are familiar with the "hardening" of cultivated plants in preparation for winter. Insects go through a similar process prior to overwintering. Exposure to frost during the middle of summer would kill a south-central Alaskan insect just as it would kill an insect from a temperate region. The cold climate insects need to "harden" to prepare for winter.

Family Phloeothripidae

Thrips overwinter as adults in plant cavities such as old galls and stem scars. Their metabolism slows down in winter as they go into diapause.

Plants and fungi that feed on insects are called entomophagous.

Fly Killer Fungus
Entomophthora muscae
Spores of this fungus germinate when in contact with the body of a fly. The mycelium of this fungus finds its way into the body of the fly and penetrates an area of the brain that controls its behavior, directing the fly (like in a sci-fi movie) to crawl up high on the vegetation. The fungus soon kills the fly, whose body remains firmly attached to vegetation. The fungus matures and produces spores that are broadcast from this vantage point. The killer fungus is most active during the cool months when humidity is high.

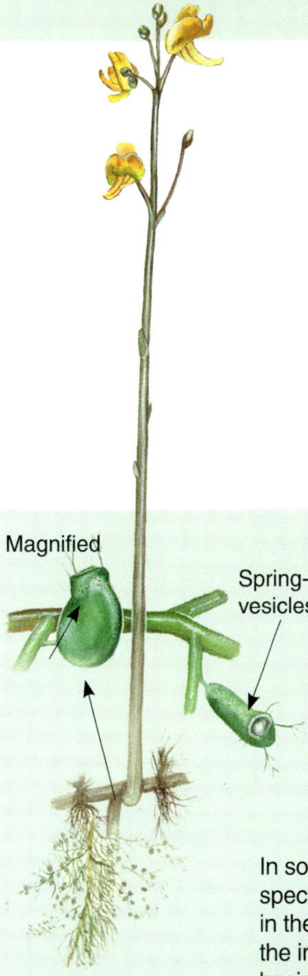

Bladderwort
Utricularia vulgaris

Butterwort
Pinguicula vulgaris

Sundew
Drosera anglica

In south-central Alaska, two *Pinguicula* and two *Drosera* species capture and digest minute insects that get tangled in the sticky coating of the leaves. The leaves curl up around the insects and digestive secretions reduce the victim to basic nutrients that can be absorbed by the plant. Three *Utricularia* species capture prey by means of submerged spring-loaded vesicles.

Insect Guilds

Most, if not, all insect species have well defined life habits. Their perpetuation depends on predictable behaviors adapted to predictable environments for each of the successive life stages. The wide diversity of behavior patterns observed among insects is an artifact of the vast number of species, each with a limited repertoire.

Each species occupies a niche, which is a specific function within the environment All insects that occupy similar generalized "niches" are grouped into guilds.

Terrestrial Insect Guilds

Herbivores		Carnivores	Scavengers
Leaf Chewers	Leaf Miners	Predators	Carrion Insects
Gall Inducers	Sap Suckers	Parasitoids	Dung Insects
Wood and Bark Borers	Pollinators	Vertebrate Blood Feeders	
Leaf Rollers	Omnivores	Parasites	

Aquatic Insect Guilds See p. 65

Collectors / Filterers	Predators / Piercers	Shredders
Collectors / Gatherers	Predators / Engulfers	
Scrapers / Grazers		

Often insects, especially those that undergo complete metamorphosis, fit into several guilds during their lifetime. For example, caterpillars (larvae) are leaf chewers but butterflies (adults) are pollinators; adult dragonflies are airborne predators while their larvae are aquatic predators. Adult *Tenthredo* sawflies visit cow parsnip flowers where they feed both on the nectar and on other insects attracted by the nectar. Aquatic collector-filterers tend to be omnivorous, consuming both animal and vegetal microorganisms.

Terrestrial Herbivores

Herbivorous species of insects are the primary feeders. Some important species are keystones for whole insect communities. That is, an array of species depend on those keystone species for their own existence. Herbivorous insect populations are normally kept in check by a combination of plant defense mechanisms, weather, predation, and parasitism.

Leaf Chewers

This group of insects feeds externally on plant foliage. Most are generalist herbivores including caterpillars, grasshoppers and some leaf beetles. Generalist herbivores are opportunistic, feeding on unrelated plants. Plant antiherbivory defense is the most efficient protection against this group of herbivores. Generalist browsers tend to be fairly common in human-disturbed sites.

Other leaf browsers such as sawfly larvae and many leaf beetles are specialists, feeding only on a few plant species, which are often toxic to generalist herbivores. These specialist herbivores have evolved resistance to the plant chemical defense and thus have access to a food source out of reach for the generalists.

Flea Beetle
Phyllotreta zimmermanni
Adult (generalist)

Grasshopper
***Melanoplus* sp.**
Adult (generalist)

Larch Sawfly
Pristiphora erichsonii
Larva (specialist)

Tussock Moth
Orgyia antiqua
Larva (generalist)

Family Cimbicid
Cimbex americana
Larva (generalist)

Willow Leaf Beetle larva
***Chrysomela* sp.**
Larva (generalist)

Sap Suckers

Sap sucking is the specialty of insects in the orders Hemiptera and Homoptera. This includes the aphids, shield bugs, plant bugs, spittle bugs, jumping plant lice and leaf hoppers. All of these insects have mouthparts modified into a stylet for tapping into the plant tissue in order to siphon sap.

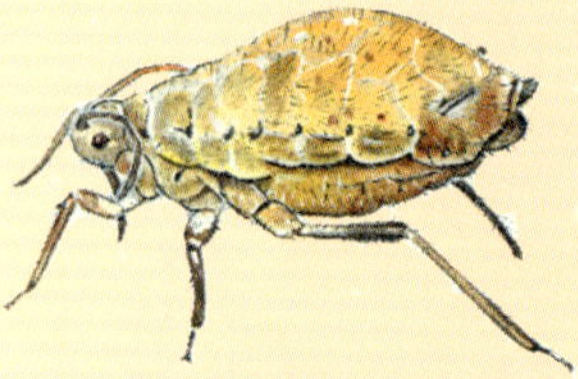

Aphid
Family Aphididae

Shield Bug
Elasmosthetus interstinctus

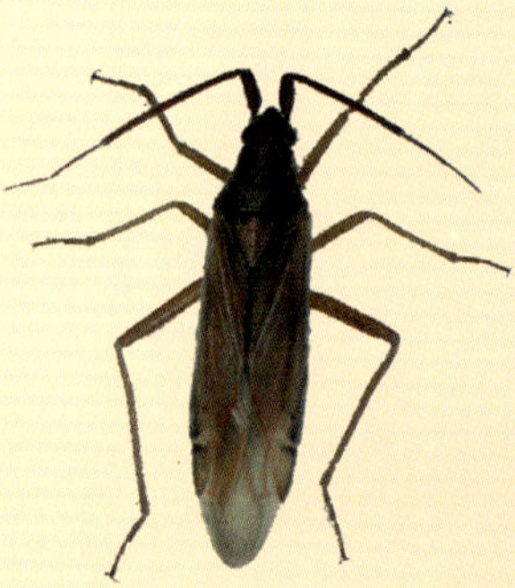

Plant Bug
Stenodema vicina

Psyllid
Cacopsylla alaskensis

Leafhopper
Thamnotethix confinis

Gall Inducers

Gall-inducing insects alter the growth of a plant, creating a novel structure. This is accomplished either mechanically, when a specific region of a plant organ is fed upon and prevented from developing, thus creating a deformity of surrounding tissue, or biologically, when deep structural changes are induced at the cellular level of the plant tissue, creating a new organ-like structure on the plant.

Gall insects are specialists, inducing galls only on a single plant species or closely related plant species. Plant galls are very consistent in their structure and in most cases, allow the identification of the gall inducer at a glance. In south-central Alaska, galls are induced on a number of plants by gall midges, cynipid wasps, aphids and sawflies.

Aphid
Tamalia coweni
Feeding by this aphid causes the leaves of kinnikinnick to curl. Unlike most aphids, *Tamalia* remains on the same plant host all year and has no asexual phase.

Aphid

Gall cross section

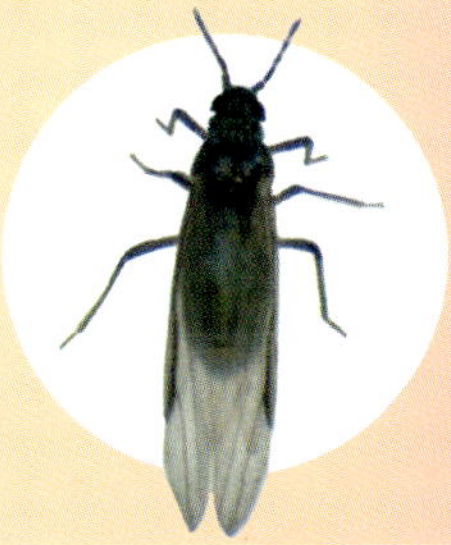

Gall Aphid
***Pemphigus* spp.**
induces galls on cottonwood leaves and petioles.

Gall Midge
***Mayetiola* sp.**
deforms the inflorescence of bluejoint reedgrass.

Gall Inducers (continued)

***Mompha* spp.** The moth caterpillars live inside the hollow stem swelling of fireweed.

Cynipid Wasp
Diplolepis polita induces prickly spherical galls on wild rose leaves.

Photo Pamela Houston

Gall Midge-induced Galls on Willows

Gall midges of the genus *Rabdophaga* induce different types of galls on willow stems and buds while the genus *Iteomyia* affects the willow leaves.

Rabdophaga rosaria "rosette" gall on undergreen willow (1X)

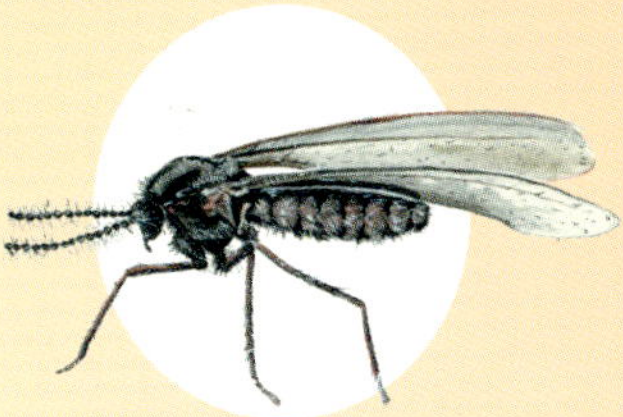

Gall Midge adult *Rabdophaga rosaria*

Rabdophaga rigidae "beaked gall" on grey-leaf willow

Gall midge ***Iteomyia* sp.** gall on the underside of Sitka willow leaf (2X)

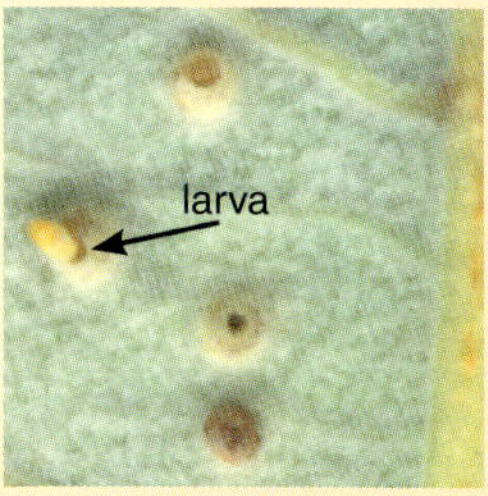

Gall midge ***Iteomyia* sp.** gall on the underside of feltleaf willow leaf (2X)

Rabdophaga salicis stem galls on Barclay willow (1X)

Gall Inducers (continued)

Sawfly-induced Galls on Willows

Various sawflies induce galls on willows. Sawflies of the genus *Pontania* induce spherical or bean-shaped galls on willow leaves. The genus has not been examined in detail in Alaska however, in western Europe where it has been well studied, at least forty-three species have been described. The majority of the species of *Pontania* are restricted to a single willow species; only a few *Pontania* species create galls on several closely related species of willows.

Pontania species can be organized in species-groups according to the gall structures. Several *Pontania* species of distinct species-groups are sometimes noted on a single willow species. A cursory survey in south-central Alaska found at least twenty species of *Pontania,* of which only a few have been described.

Sawfly *Euura* sp.

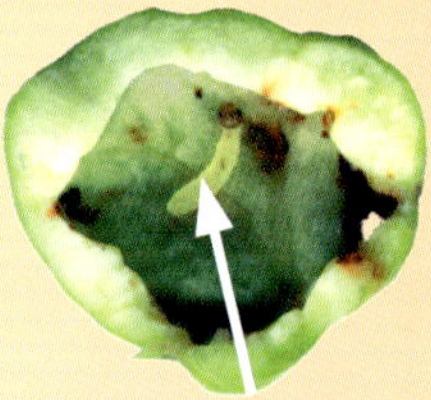

Pontania larva inside the gall

Pontania* sp. *dolichura *group* gall on diamond-leaf

Pontania* sp. *proxima *group* gall on Barclay willow leaf

Pontania reticulatae viminalis *group* gall on net-vein willow

***Pontania* sp.?** Gall on Barclay willow leaf

***Pontania* sp.?** Gall on Barclay willow leaf

***Euura* sp.** gall on Scouler willow

Sawflies of the genus *Euura* induce galls on woody willow stems.

Leaf Miners

The larvae of leaf miners feed on the internal spongy layer of leaves without damaging the outer layers. The pattern of the galleries is characteristic for each leaf miner species. Immature leaf miners are out of the reach of most generalist predators but are targeted by specialized parasitic wasps and predators.

Aspen Leaf Miner
Phyllocnistis populiella
During outbreaks, high densities are reached, with several minute (3/16" long) caterpillars of the aspen leaf miner *Phyllocnistis populiella* mining every single aspen leaf. The combined mines delaminate the whole surface of aspen leaves. The leaves, lime green when healthy, turn silver by mid-summer before turning brown and falling off prematurely. The minute moths emerge in late June or early July in the early evening, forming clouds that appear as silvery ice crystals. Such outbreaks have occurred over extensive areas for the last few years. In 2004, millions of acres were affected in interior Alaska as far south as Wasilla.

Fenusa pusilla and *Profenusa thomsoni* are two species of minute sawflies whose larvae mine birch leaves.

Profenusa thomsoni, the amber-marked leaf miner and *Fenusa pusilla,* the amber leaf miner are two minute sawflies whose larvae mine birch leaves, causing large brown blotches. In Alaska, the heavy infestation by *P. thomsoni* spreads over thousand of acres from Anchorage to Fairbanks.

Fenusa pusilla adult

Photos U. of Guelph

Leaf blotches caused by ***Fenusa pusilla*** larvae

Leaf Rollers

Damage to the leaf cells on a single face of a developing leaf blade causes the leaf to curl up. Several leaf browsers use the rolled leaf as a shelter. Some species use this roll only in the early larval stage. Others, like *Phyllocolpa* spp. complete their larval development inside the shelter.

Rolled birch leaf

Rolled aspen leaf

***Phyllocolpa* spp.**
This sawfly causes a section of the margin of willow leaves to curl up tightly. The larva completes its development inside the shelter. Although protected from the elements and generalist predators, *Phyllocolpa* have many enemies, mostly specialist parasitoids.

***Phyllocolpa* sp.**
on felt-leaf willow

***Phyllocolpa* sp.**
on Sitka willow

The rolled leaf margin has been opened to expose a ***Phyllocopa*** larva.

Pollinators

Insects are attracted to the sweet secretions of flower nectaries. As they feed on the nectar, they inadvertently pick up pollen which they transport to the next flowers they visit. Pollinators include well-known insects such as bees, bumblebees, butterflies, moths, and also long horned beetles and countless flies, beetles, and wasps.

Swallowtail
Papilio machaon

Long-horned Beetle
Judolia montivagans

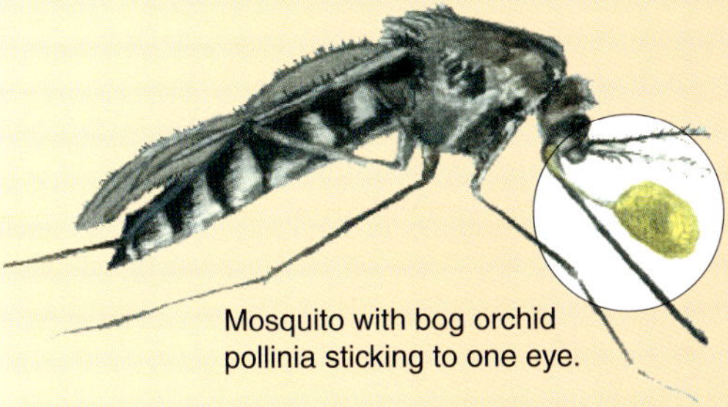

Mosquito with bog orchid pollinia sticking to one eye.

Fungus gnat with an orchid pollinia glued to the eye.

While most flower pollen is dispensed in loose grains, orchid pollen is clustered into a pair of yellow masses called pollinia. The pollinia mass extends into an arm that protrudes in front of the nectaries. The end is sticky and adheres to the forehead or the eye of a visiting insect which then flies with the pollinia held fast in front of its face. While the insect visits the next orchid, the pollinia mass rubs against the pistil, to which the pollen grains adhere. Mosquitoes, fungus gnats and bees have been observed flying with orchid pollinia glued to their faces.

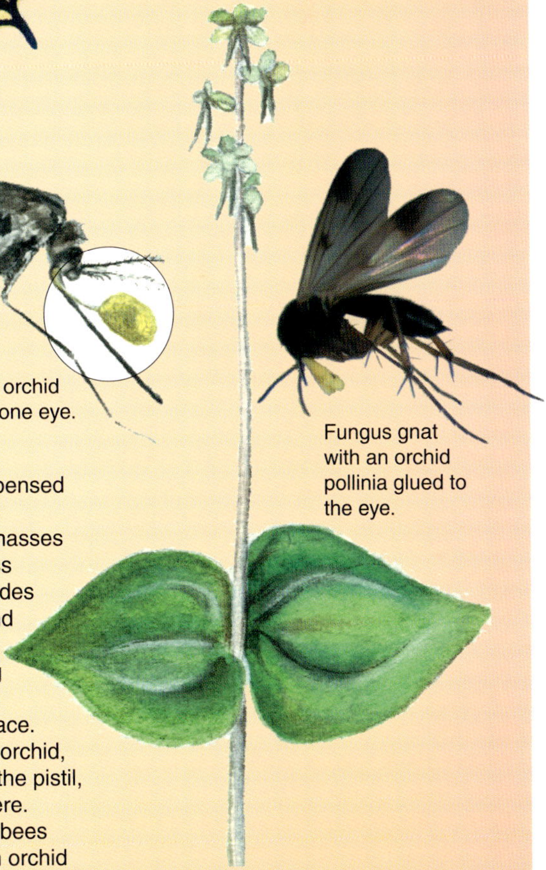

Twayblade Orchid

Wood Borers and Bark Beetles

In a tree, most of the nutrients available for herbivores are piped down from the foliage to the roots through the phloem, the spongy layer between the outer sapwood and bark and the sapwood. Most insect borers are concentrated in that nutritious layer. The few other borer larvae that develop deep inside the sapwood and heartwood typically are slow to develop, often taking several years to mature. Some of the deep borers actually do not feed directly on the wood, but graze on fungi "mold" filling their galleries. These symbiotic fungi are injected by the female wood borer at the time of egg deposition into the tree.

Spruce bark beetle larvae bore the phloem layer, between the wood and the outer layer of the bark.

Dendroctonus rufipennis

***Saperda* sp.**
Willow stem borer pupa in 2-year-old feltleaf willow twig.

***Trypodendron* sp.**

Ambrosia beetles have a symbiotic relationship with a blue stain fungus. The fungus is carried to the tree by the beetle at the time the galleries are bored into the wood. The fungus attacks the sapwood and mold fills the galleries, providing nutrients for the beetle larvae.

Oblique tunnel entrance in the middle of the chamber in the phloem layer, with bark removed for clarity. (white spruce)

Larvae of spotted sawyers first excavate a chamber roughly 2" in diameter below the bark. They then bore an oblique tunnel deep into the wood.

Monochamus scutellatus

Carnivores

Predators hunt their prey and eat them or feed them to their brood. In the course of their lives, they kill a lot of prey and are often opportunistic, attacking a number of unrelated species. Predators are found in the orders Hymenoptera, Diptera, Coleoptera, Odonata, and Neuroptera.

Parasitoids complete their larval development on or inside a single individual host, nearly always killing it. Each species of parasitoid is usually intimately associated only with a single or a few closely related host species. Larval parasitoids feed on a single host and are always smaller than their victims.

Predators

Cicindela oregona
The predatory tiger beetle larva awaits its prey, hiding at the entrance of a vertical shaft dug in clay. The head and thorax are modified to cover the shaft opening. When prey passes, the larva swiftly erupts from the hole, pounces on its victim, and drags it back into the hole.

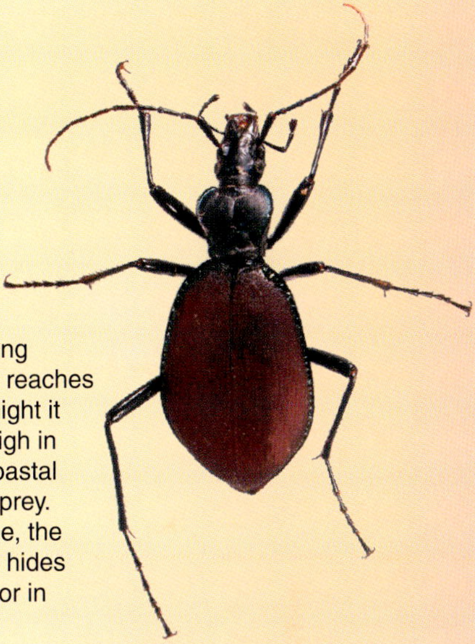

This snail-eating ground beetle reaches 3/4" long. At night it climbs fairly high in trees of the coastal forest to hunt prey. During daytime, the ground beetle hides under stones or in the duff.

Scaphinotus angusticollis

***Anthocoridae* nymph**
The nymph of this species of minute pirate bug lives in crevices of tree bark and kills many types of prey. Here it is seen sucking a spruce bark beetle larva dry. Normally, the bark beetle spends its life in galleries under the bark. But if exposed by damaged bark, it become easy prey for a predator such as this bug.

Parasites

No Alaskan insects are known to be human internal parasites. Botflies and warble flies parasitize caribou and reindeer, however. Botfly larvae develop in the back of the animals' throats and the warble flies develop on the rump, just under the skin. Adults of these flies resemble hairy bees, but do not have functional mouths and do not feed in the adult stage.

External parasites and blood feeders can be vectors of blood-borne diseases.

Lousefly
Ornithomyia

Louseflies parasitize a number of bird species in Alaska. They feed only as adults. The females give birth to fully-grown larvae ready to pupate.

Flea

Fleas live on their hosts only when adults. As larvae, they live in debris near their host's usual resting place: nest, den or house. Fleas are specialists and live only on a few host species, although they occasionally bite other species.

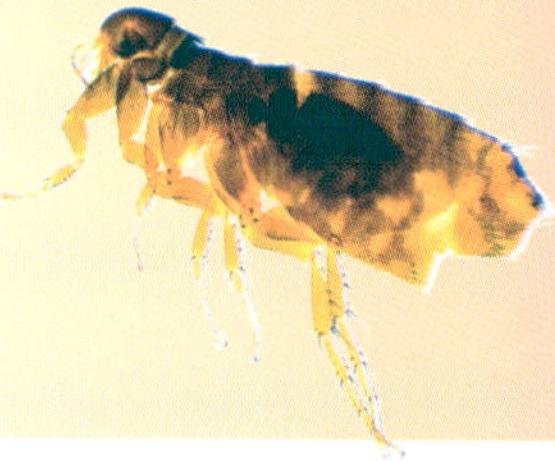

Louse
Pedicularis humanus

Lice are specialists and spend their whole life cycle on their hosts.

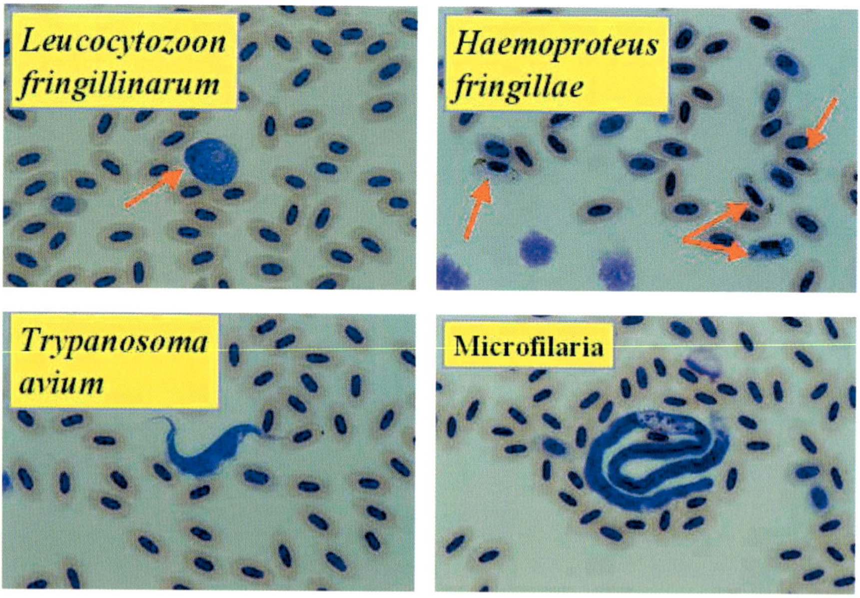

These microorganisms are blood parasites transmitted by blood feeding-insects such as the lousefly. Photo courtesy of Pierre Deviche

Parasitoids

Most parasitoids are specialists whose larvae develop on a single or a few closely related species of hosts. This characteristic makes them attractive for use in bio-control because specific pest species can be targeted. One to more than a thousand parasitoid larvae may develop on a single host.

The life cycle of the parasitoid is often closely synchronized with that of the host as the inoculation by the female parasitoid must happen at a specific life stage of the host, usually egg or larva. Host larvae that develop concealed in plant tissue are typically targeted either at their egg stage or early larval stage, before they are hidden deep in the plant tissue

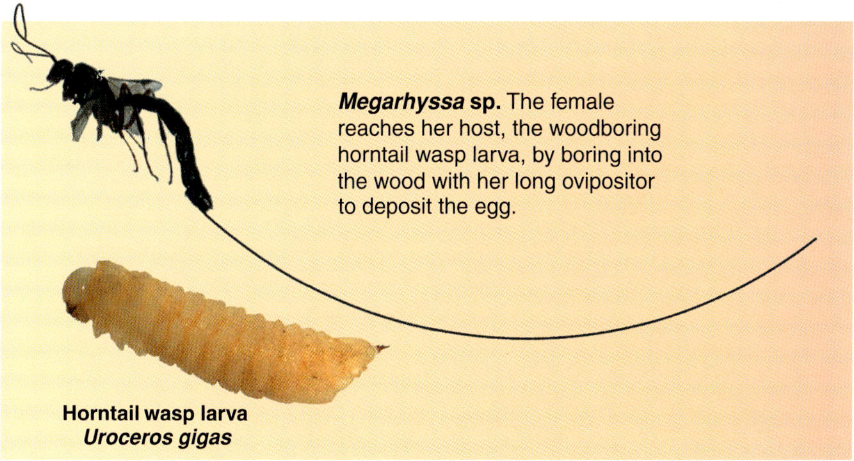

***Megarhyssa* sp.** The female reaches her host, the woodboring horntail wasp larva, by boring into the wood with her long ovipositor to deposit the egg.

Horntail wasp larva
Uroceros gigas

Parasitoid larvae develop either externally on the host (ectoparasitoid) or internally in the host (endoparasitoid).

Ectoparasitoids

The larvae develop outside the body of their host. If the host lives in the open, the eggs are either attached to the body of the host or are inserted partly inside the host's body. If the host is living inside a plant cavity, the egg may be deposited near the host. The parasitoid larva feeds externally on the host that has been paralyzed either by the female at the time the egg was laid or by the larva at the first feeding.

Endoparasitoids

Female endoparasitoids insert their eggs inside the host or leave the eggs sticking to the host, in which case, the larva enters the host body on its own. Larvae of two families of minute wasps, Mymaridae "fairy flies" and Trichogrammatidae, complete their development entirely inside a single host insect's eggs. Winged adults of these egg parasitoids may be as small as 1/100 of an inch. A few species of both families are used successfully in biological control programs, in some cases parasitizing 90% of the targeted pest species. Females of other endoparasitoids often insert their eggs inside very young larvae or even inside an embryo.

Endoparasitoid larvae postpone their own growth for several months. The host larva is allowed to grow normally for an extended period and the host is disabled only after it reaches the last larval stage or after it pupates. The parasitoids then consume the whole host and pupate either inside the empty host skin or near the host's remains.

Parasitoids (continued)

Parasitoids of adult insects inject an egg inside the adult host or the minute parasitoid larvae penetrate the host's body through the softer cuticle of a joint.

Ectoparasitoid

The ectoparasite egg is deposited near or on a mature host larva.

The young ectoparasite larva crawls and feeds externally on a paralyzed host larva.

After consuming the host, the larva matures and pupates near the host's remains.

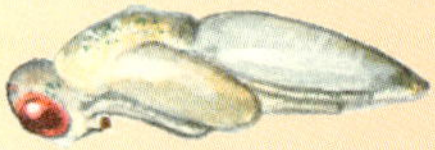

The pupa matures and the winged adult emerges.

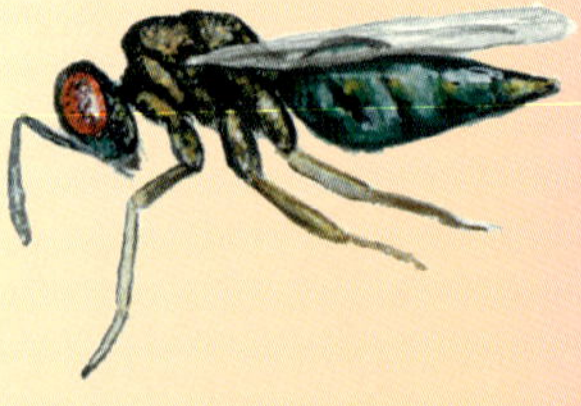

Mesopolobus diffinis
(Hymenoptera, Pteromalidae)

Endoparasitoid

The egg is deposited in the embryo or inside a very young host larva.

The young endoparasite larva delays development inside until the host larva reaches maturity.

After consuming the host, the larva matures, then emerges from the host's empty skin and pupates near the host's remains.

The pupa matures and the winged adult emerges.

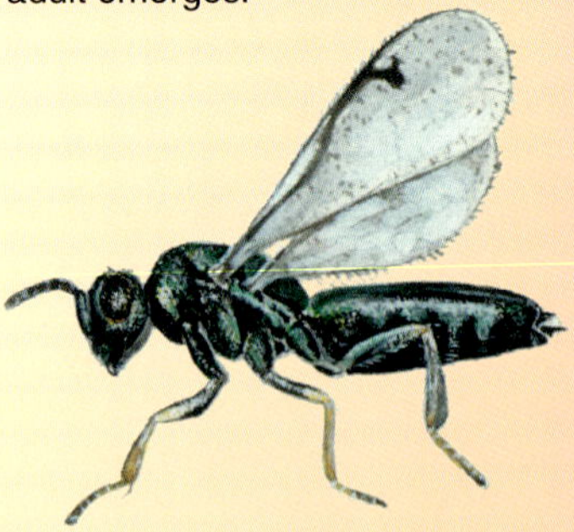

Gastrancistrus salicis
(Hymenoptera, Pteromalidae)

Life Cycle of two Parasitoids of the Willow Rosette Gall Midge

Dryinidae

Dryinid wasps parasitize leafhoppers. The female dryinid grabs the leafhopper from the back with highly modified forelegs and injects an egg into the body of the leafhopper. The larva emerges from the egg inside the host and after a few days appears partly exposed outside the body of the host. As the parasitoid develops, the outside layer of the parasitoid skin tears lengthwise but remains attached. Once mature, the larva leaves the host and pupates in the soil. The parasitoid usually does not kill the host but prevents the immature host nymph from further molting and interrupts the maturation of the sexual organs.

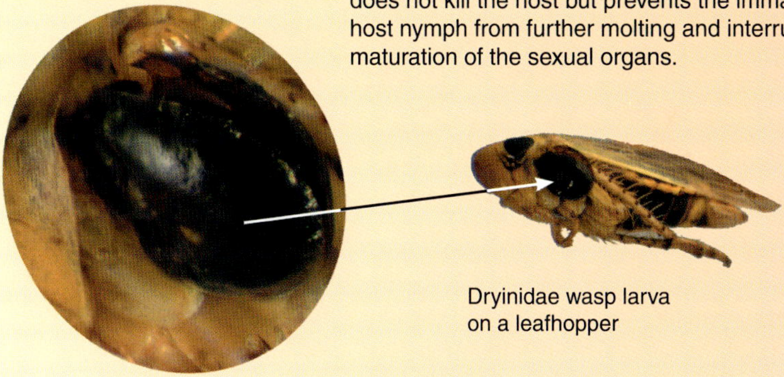

Dryinidae wasp larva on a leafhopper

Stonefly with yellow ectoparasitoid larva under the base of the left wings

Hyperparasitoids

Aphids are often parasitized by *Aphidius* (Braconidae: Hymenoptera): the female *Aphidius* injects an egg inside the body of the aphid. After about three days, the *Aphidius* larva, an endoparasitoid, emerges from the egg inside the aphid and feeds on its internal organs. After about a week, the aphid turns pale, becomes paralyzed and dies, its skin inflated like a balloon and dried like parchment. This stage is called an "aphid mummy."

Still inside the aphid mummy, the *Aphidius* larva spins a cocoon and pupates. After another week, the now adult *Aphidius* chews a hole out of the back of the aphid mummy and emerges. That is, unless a *Charips* has come along. *Charips* females search for aphid mummies containing almost mature *Aphidius* larva. When it finds such an aphid, it injects its egg into the *Aphidius* larva through the skin of the aphid. The *Charips* larva develops inside the *Aphidius* larva. Before maturing, the *Charips* wasp larva exits the *Aphidius* body and while remaining inside the aphid mummy, finishes consuming the *Aphidius* and then pupates. After a few weeks, the *Charips* cuts a hole through the skin of the aphid mummy and emerges. *Charips* are obligate parasitoids of *Aphidius,* which make them hyperparasitoids of aphids.

Dendrocerus is also a hyperparasitoid of *Aphidius* or other endoparasitoids inside aphid mummies. But the *Dendrocerus* female will lay an egg, whether or not there is already a hyperparasitoid or parasitoid such as *Charips* inside the aphid mummy. In such a case it is tertiary parasitism: it is almost like the Russian doll that contains another doll that contains another doll except that *Dendrocerus* is an ectoparasite on the last larva.

Only one adult insect will emerge from the aphid mummy and the size diminishes with each successive level of parasitism since no additional nutrients enter the food chain after the original host has stopped feeding. Hyperparasitoids can be a problem in biological control of pests because they reduce the population of beneficial parasitoids.

Empty aphid mummy
The parasitoid emerged through the hole at the rear.

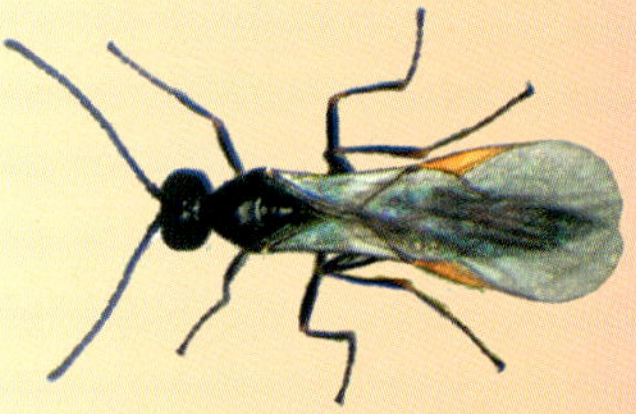

***Aphidius* sp.**

***Charips* sp.**

Dendrocerus carpenteri

Vertebrate Blood Feeders

The immature stages of most of these biting insects live in wetlands habitats. Adult females need the proteins provided by a vertebrate blood meal to mature their eggs. The sheer number of mosquitoes or black fl ies renders most wetlands inhospitable to humans during the warm season.

Horse Fly
Hybomitra sp.
Family
The loud humming sound of this powerful flier can send dogs in a panic. The larvae live in wetlands.

Deer Fly
Chrysops excitans
Tabe hidao
The eyes of the live adult are a green or yellow-metallic color. The larvae live in decaying vegetation in damp soil.

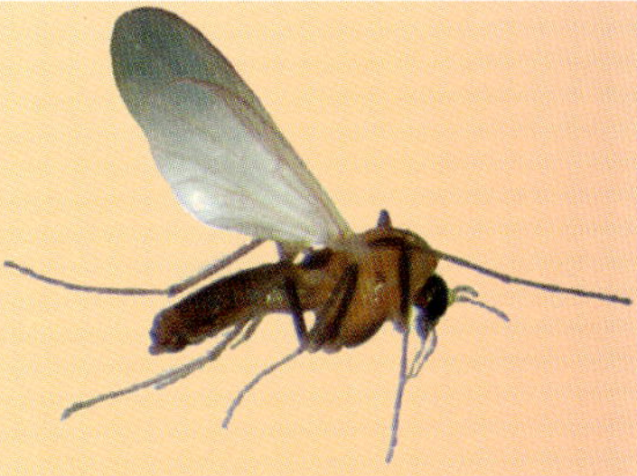

No-see-um
Family Ceratopogonidae
This minute biting fly is small enough to penetrate through mosquito netting. The larvae live in saturated soil.

Moth Fly
***Duckhousiella* sp.**
Family Psychodidao
The larvae live in moist organic matter.

Black Fly
Family Simuliidae
The black fly larvae live fastened to rocks on the bottom of fast-moving current and can be extremely abundant. The bite of the adult can cause hard welts and itching.

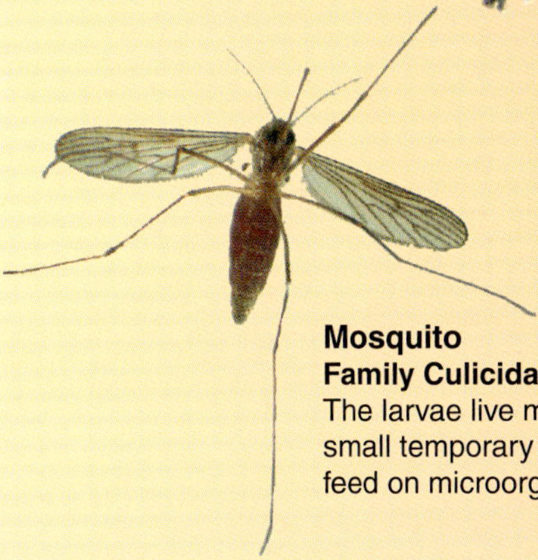

Mosquito
Family Culicidae
The larvae live mostly in small temporary pools and feed on microorganisms.

Scavengers

Carrion Insects

The colorful burying beetles prefer small carcasses, the size of small birds or voles, which they bury and store to feed their larvae. Carrion beetles always carry a number of orange mites clinging to the underside or the rear of their bodies. These mites and the beetles sometimes have a mutualistic relationship, helping each other survive. The beetle transports the mites to the food source and the mites eat fly eggs, thus suppressing maggots that otherwise would compete with the carrion beetle larvae.

Scarabids

Like carrion insects, dung insects fly some distance to reach dung piles. Coprophagous insects are mostly scarabs and flies. The scarabid *Aegialia* is fairly common on moose dung and in cow patties. These insects are very important in the recycling of nutrients in the food chain. By burying the dung, nitrogen is incorporated into the soil instead of being released into the atmosphere as ammonia.

Insects can be classified as generalists or specialists. Generalists, whether herbivores or carnivores, are opportunistic in nature and eploit whatever suitable resource is available. Their diet varies with the availability of resources. Generalists are able to take advantage of the most abundant resources, within certain constraints, which can be defined in general terms such as ecology of the habitat, size of prey, etc. Chemical or other defense mechanisms often deter generalists.

Specialist insects have a more limited range of behavior and usually fill a very precise niche within an insect association. Parasitoids, inquilines (See p. 59) and specialized insects associated with any herbivorous insect species constitute that herbivore's insect community. The herbivorous species is the keystone of this assemblage of sometimes a dozen or more insects species. Usually several communities live side by side on any given plant. All herbivores compete for the plant resource but direct interactions between other species of the two discrete insect communities are limited. Within the insect community, each insect occupies a precise niche, according to its species, sex, age etc., interacting primarily within its own association, oblivious to insects from other communities.

For instance, a typical Barclay willow shrub within a road right-of-way on the Kenai Peninsula will possibly have the following herbivores: willow leaf beetles, rosette gall-inducing gall midges, aphids, gall-inducing sawflies, and leafhoppers. Relationships between specialist insects associated with the willow gall midge and those associated with aphid parasitoids are minimal. There is little direct interaction between two of these organisms besides attempting to occupy the same physical space or both being eaten by the same generalist predator, such as a dragonfly or a chickadee.

The gall midge remains isolated from generalist predators except for the brief flight period as a flying adult, looking for a mate and for a site to lay its eggs. Only opportunistic predators or larger herbivores such as a moose may unite the aphid parasitoid and the gall midge in their fate. In the following pages, we will look at communities common in south-central Alaska and some of the dynamics involved. With each community we will look at a different aspect of the interactions: chemical defense by willow leaf beetles against predators, population dynamics of insects associated with aphids, the intricate relationship among the main herbivore a bark beetle and its associates and closely related herbivore insects with associations with gall midge-induced galls on Barclay willow.

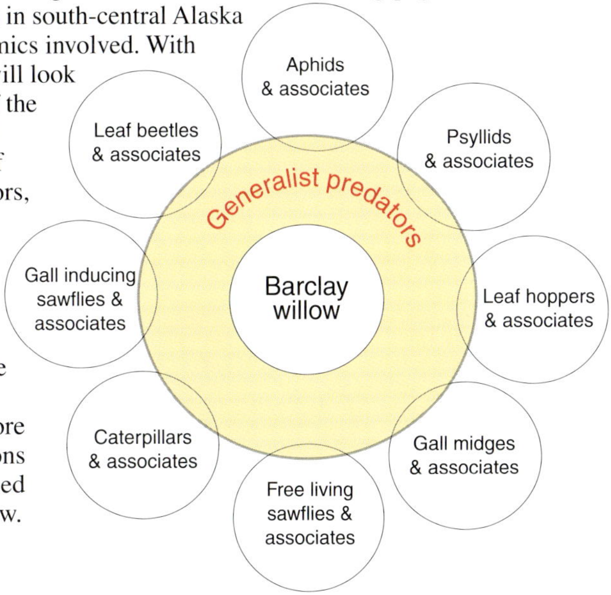

Insects Associated with Willow Leaf Beetles, Chrysomela falsa

Life Cycle

Willow leaf beetles overwinter as adults, hidden in the duff under shrubs. In the spring they emerge and gather on their food plants where they mate. The females lay elongated, shiny, yellow eggs in compact groups sticking to the new leaves. After a few days, the larvae emerge and start feeding. The larvae are black and their elongated bodies, with laterally developed legs, are reminiscent of lady beetle larvae. The larvae are voracious and soon skeletonize sections of the center of the leaf, while the adults tend to feed on the margins. When the larvae are fully grown, they fasten themselves to the surface of a leaf by the end of their abdomens and pupate in that position. After a few weeks, the adults emerge from the pupae and feed on the foliage. The adult elytra display a brown and black pattern at first, the bright red color developing slowly.

Willow Leaf Beetle Life Cycle

The fact that these leaf beetles live in the open, on the upper face of the leaves in broad daylight, hints that they are somehow protected from generalist predators.

Chemical Defense

Willows and poplars synthesize chemicals that deter most herbivores. One of these chemicals is salycilin, from which aspirin was derived. For all its medical virtues, large amounts of aspirin ingested on an empty stomach cause great discomfort to humans, and hence it also deters generalist herbivore insects. A few herbivores are able to overcome this chemical defense and feed on willows with impunity and some even thrive on the plants. Moose are one example and willow leaf beetles are another. Not only does the beetle overcome the chemical defense of the plant, but it sequesters the salycilin and modifies it into salicylaldehyde. This powerful chemical can be discerned as a pungent odor when a colony of *Chrysomela falsa* larvae feeding on Barclay willows is disturbed. Droplets of the white volatile chemical pearls appear on the tips of protruding glands aligned on either side of the abdomen.

Willow leaf beetle larva dispensing the volatile repellent

Willow leaf beetle eggs are also protected by the same chemical. Each egg contains enough toxin to kill an ant. The pupae, however, appear to be chemically unprotected as phorid flies and their parasitoids, alysiine braconid wasps,[1] lay eggs inside pupae.

Not all willows synthesize salicylin, and leaf beetles synthesizing salicylaldehyde are protected only when they are feeding on salycilin-rich willows. Barclay willow and cottonwood synthesize salycilin, whereas Sitka willow does not. Leaf beetles that are not chemically protected lead discrete solitary lives because they are vulnerable to predatory bugs, lady beetles, ants, hoverfly larvae, various egg parasites, etc. Still, a few predators and parasitoids have been able to overcome the chemical defense of leaf beetle larvae and a hoverfly, *Parasyrphus* sp. is even attracted to chemicals released by distressed larvae and uses them to locate its victims!

1 Thus *Chrysomela* hyperparasite See p. 48

Pacific Willow
Leaf skeletonized by willow leaf beetle larvae

Insects Associated with Aphids

Typically, aphids overwinter on their primary host plant, as eggs sticking to a branch or in the crevice of tree bark. Early in the spring, as soon as they emerge, the small nymphs tap into the abundant sap flowing within the tree. Without prior mating, these wingless females (fundatrices) soon give birth to live young, also wingless females. These multiplying aphids soon form dense colonies on new shoots, doubling in number every other day. When the foliage of the host matures, the plant's need for nutrients decreases and the nutritional quality of the plant sap declines, forcing the aphids to migrate to a secondary plant host, often a grass or forb where they spend the rest of the summer. It is often during this mass migration that aphids suddenly appear in great numbers on cultivated plants. When autumn comes, a sexual generation of males and females develops. These migrate back to the primary host, on which they lay the eggs that overwinter.

Dense colony of aphids on fireweed, a secondary host plant.

The tremendous reproductive potential of aphids is balanced by their vulnerabilities. Aphids have poor defenses against predators. Their body is not protected by a strong exoskeleton and when feeding, the aphid's stylet is deeply implanted in plant tissue which prevents it from making a quick escape, leaving it as vulnerable as a sitting duck. Such a convenient source of protein does not go untapped. Scores of insectivorous birds and invertebrate generalist predators such as paper wasps include aphids in their diet. Adults and larvae of several species of lady beetles and lacewings, larvae of flower flies and gall midges are specialized predators of aphids and consume little else. Numerous species of parasitoids, including one entire family of parasitic wasps, Aphelinidae, and one subfamily of wasps, Aphidiinae, use exclusively aphids as their hosts. Both predators and parasitoids have a slower initial reproductive rate and are only able to have an impact on their host-prey population late in the season.

By then the aphid population has built up to large numbers and predators are literally surrounded by their quarry.

Predators may be slower to develop than aphids but they are voracious. Brown lacewings consume about 40 aphids in their lifetime and lay some 460 eggs. This makes them a good candidate for the biological control of aphids.

Maggot-like flower fly larvae will raid aphid colonies. One of these larvae can eat up to 1200 aphids during the course of its development and often consumes entire aphid colonies.

Ants, which obtain honeydew from aphids by stroking them with their antennae, protect aphids by chasing off predators. Predatory flower fly larvae defend themselves by spitting a sticky substance on the ant which keeps it distracted.

Besides the mass migration from primary to secondary plant host, aphids do little to escape predation.

When the population of aphids peaks, the predators, such as orange gall midge larvae, are surrounded by their aphid prey.

Insects Associated with Aphids (continued)

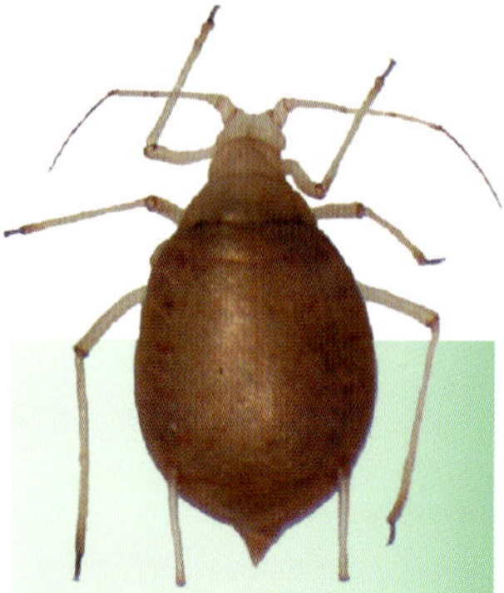

Aphid mummy
This aphid has been killed. The insides have been eaten away by a parasitic wasp, such as *Aphidius*. All that remains of the aphid is a hollow inflated skin. The predator is still inside, in the pupal stage.

Green Lacewing
Both adults and larvae are aphid predators. They are often sold in greenhouses.

Gall Midge larva
The name of the family "gall midge" does not accurately describe the biology of the many insects within this goup that do not induce galls. Instead they are predators or fungus eaters. The species illustrated here is a voracious predator of aphids.

Aphidius

Two subfamilies of small parasitic wasps, Aphidiinae and Aphelinae, are exclusively endoparasites of aphids.

Asaphinae
Asaphes species are hyperparasites of aphids through Aphidiidae. They are shiny metallic green.

To effectively control aphids, numerous parasitoids or predators have to be introduced early in aphid colonization in order to prevent the "fundatrix," or original prolific wingless females, from establishing new colonies. This needs to occur, either in the spring when they attack the primary host or, later, when they migrate to the secondary host. This can be done by saturating the primary host plant in the spring or the secondary host plant in midsummer with predators purchased for this purpose, often lady beetles or lacewings. The predators need to be reintroduced every year, because even if they were to survive the winter, they would still lag behind the aphid buildup. Also, since the predators are released in great numbers to overwhelm the aphid fundatrix, they soon run out of prey and are forced to disperse or perish.

Honeydew

Plant sap is rich in carbohydrates (sugars) and relatively poor in other nutrients. Aphids have to ingest large amounts of sap to fulfill their nutritional needs. Excess sugars pass unmodified through their digestive tract and are forcefully expelled through the cornicles (tubes sticking out of the rear of the abdomen). This sugar-rich fluid, called honeydew, coats the foliage of shrubs and automobiles below large aphid populations. The honeydew is strongly sought by many insects craving sugars. Butterflies, moths, flies, beetles and many other insects lick the honeydew residue on leaves. Aphid colonies are visited by these insects for their sugar as often as are flowers. Ants in particular have developed a mutualistic (beneficial to both) relationship with aphids. They have often been compared to shepherds tending their flocks. Ants collect the honeydew secreted by the aphids and in return, protect the aphids from their enemies.

Flower Fly larva
The larvae are solitary and common among aphid colonies. Their color varies from dark brown to grey, pale brown or green.

Lady Beetle
Both adults and larvae are voracious aphid predators.

Insects Associated with Rosette Galls on Willows

The bunched flowers of willow form a catkin.

Perhaps you have noticed on bushy willows growing along the road right-of-ways strange "roses," green in the summer, reddish in the fall and persisting blackish in the winter. These are not willow flowers but are galls induced by a gall midge of the *Rabdophaga rosaria* species-group. The true willow flowers are minute and bunched into catkins and occur only on female plants. Early in the spring, before willow leaves begin to open, the female *Rabdophaga rosaria* lays a few red eggs near the tip of the winter bud. One of the midge larvae emerging from an egg penetrates the center of the willow bud. It positions itself upside down, with the mandibles touching the shoot's meristem and starts feeding on the tip, preventing the new shoot from elongating. The leaves grow but remain clustered around the tip of the supressed stem, resembling a Brussels sprout. The single larva lives in the center while the gall expands around it and remains there until the following spring when it pupates, and then emerges as a midge.

The larva is a barrel-shaped orange grub about 3/16" long. The adult midge of the same length is short lived, a few days at the most.

In the autumn, the decline of chlorophyll in gall-leaves reveals the background color of the plant tissue, which is red.

The gall midge larva lives in a cavity in the middle of the gall.

Enemies

The longest part of the gall midge lifecycle is spent inside the fully developed gall, quite safe from generalist predators. Predation by generalist predators is limited to the short period during which the egg or the newly emerged larva is exposed.

All of the larvae of gall midge parasitoids develop on the fully grown gall midge larva, when the host reaches maximum size.

Most of the casualties endured by the gall midges are inflicted by specialized parasitoids. Internal parasitoids such as *Platygaster* sp. and *Gastrancistrus salicis* inoculate early in the cycle and defer their development until that ideal last larval stage.

External parasitoids such as *Torymus cecidomyiae* and *Mesopolobus diffinis* attack their host later in the cycle, when the last instar larva of the host midge is fully developed and hidden inside the gall. The eggs are deposited in the vicinity of the host larva and the parasitoid larvae actively seek the host. The timing of the parasitoids is finely tuned to match that of their hosts, which is made possible because of the predictability of the life cycle of the gall midge.

Fully grown larva of the gall midge *Rabdophaga rosaria*

Parasitoids

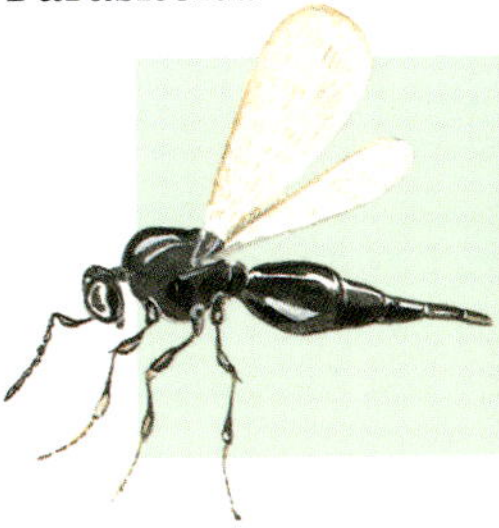

The endoparasite *Platygaster* inoculates the host when the gall midge is still an embryo inside the exposed egg.

Gastrancistrus salicis, an endoparasitoid, inoculates the newly hatched larva before it gets out of reach behind the scales of the rose gall.

The external parasitoid *Torymus fuscipes* female is equipped with an extra long ovipositor that allows it to bore into fully developed galls and insert the egg close to the hidden host.

Mesopolobus diffinis has an ovipositor with a short reach and cannot contact the host larva directly in the large gall. The parasitoid larva has to seek the host on its own by crawling between the scales of the gall.

Other Associates

Inquilines are insects that live inside galls induced by another insect. They are totally dependant on the other species to create this microhabitat where their immature stages develop. Often, but not always, inquilines belong to the same family as the gall inducer. Some inquilines modify the structure of the original galls. That may or may not cause the premature death of the original gall maker.

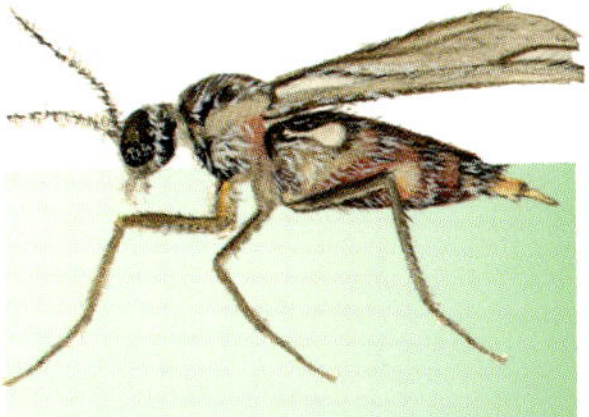

Small inquiline gall midge larvae of the genus *Dasineura* depend on the presence of rosette galls to live. Their larvae develop and pupate at the base of the scales of the rosette.

The larva of the sawfly of the genus *Euura* sometimes finds its way into the gall from which it feeds on the plant tissue, often killing the larva of the gall-inducer as it bores though the gall.

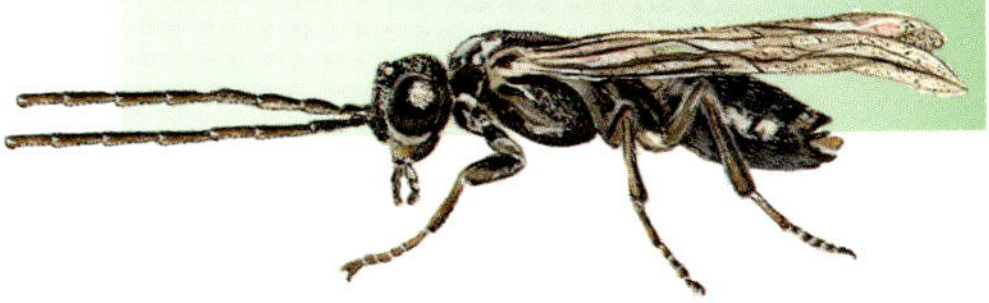

Insects Associated with Spruce Beetles

If you peel off the bark of a spruce tree dying from a massive attack by spruce bark beetles at the end of summer or early in the spring, you will find mostly immature insects. As in the majority of food webs dominated by insects, most of the predation and parasitism occurs when both the prey (or host) and the predator (or parasitoid) are in their larval stage. It is, after all, during the larval stages that insects accumulate the necessary nutrients required to carry them through the remainder of their life cycle. Most of the larvae under the bark are quite distinctive and can readily be identified to the family level by comparing them with the illustrations provided here. Most of the species illustrated here are specialists and are usually not found associated with other insects. Ectoparasitoids can be found either on paralyzed hosts or near the remains of hosts. Endoparasitoids will only be noticed at a later stage when the host larva is discolored or looks abnormal.

Spruce Beetle
Dendroctonus rufipennis

The larvae have a large brown hardened head armed with black mandibles. The cream-colored body is wrinkled and translucent. The larvae live in chambers excavated in the inner bark.

Spear-winged Fly
***Lonchaea* sp.**

The larvae are predatory on bark beetles, and are often found under the bark of spruce beetle-infested trees.

Bark-gnawing Beetle
Family *Trogositidae*

The adults are predatory on adult small-sized bark beetles.

Cylindrical Bark Beetle
Lasconotus borealis

The larvae feed on bark beetle larvae.

Darkling Beetle
Corticeus praetermissus
The larvae resemble meal worms with their wiry body, light brown rigid segmented abdomen, two short stubby antennae, and short articulated legs at the front of the body. They are occasional predators of bark beetle larvae. (adult illustrated p. 124).

Larva

Roptocerus xylophagorum
The front of the larva is broad and truncated, the end of the abdomen pointed, whitish-translucent, and glossy. The larva parasitizes bark beetle larvae while they are hidden in their chambers. The female parasitoid uses chemicals released by the frass, a mixture of fecal material and boring dust, created by the spruce beetle, to locate the entry holes to the spruce beetle galleries that containing immature beetles. The female parasitoid reaches her host, the fully grown beetle larvae, either by crawling into the galleries or by drilling through the thin bark with her ovipositor.

Lateral view Dorsal view

The larva is ectoparasitic on bark beetle larvae. The adults emerge in great numbers from logs infested with spruce bark beetles.

Pteromalid Wasp
Dinostichus thompsoni

Rhizophagid Beetle
Rhizophagus dimidiatus
Feeds under the bark, on eggs and immature bark beetles, especially *Ips.*

Checkered Beetle
Thanasimus hubilus
This clerid beetle is a predator of spruce bark beetles and overwinters either as larva, pupa or adult. Very common in the Alaskan interior, and found in small numbers on the Kenai Peninsula.

Insects Associated with Spruce Beetles (continued)

Spruce beetle galleries are tightly clogged with debris. The female braconid ectoparasitoid *Coeloides* sp. reaches its host, the mature larva of the spruce beetle, by ovipositing directly through the bark into the individual larval chamber. The parasitoid larva consumes the bark beetle larva during the warm summer season and by fall, all that remains of the parasitized spruce beetle larva is the head capsule near the parasitoid's cocoon. The parasitic wasp emerges in the spring after cutting a circular exit hole through the bark.

***Coeloides* sp.**

Underside of spruce bark

***Coeloides* sp.**
cocoon in bark beetle larvae the chamber

Flat Bark Beetle
Cucujus clavipes
The body of the adult is flattened, so that the beetle fits between the sapwood and the bark where bark beetles live.

Brown Spruce Borer
Tetropium cinnemopterum
The young larva first bores a tunnel the phleum and cambium, just under the bark, and after a couple of months extends the tunnel into the xylem perpendicular to the trunk. After 1" to 1-1/2" the tunnel turns parallel to the trunk for another 1" to 1-1/2". The beetle pupates in the chamber and exits the trunk by boring an oval though the bark. During the earliest part of its life, this longhorn beetle lives in the same layer as the spruce bark beetle.

Rove Beetle
Staphylinidae
Predator on bark beetle larvae

Long-legged Fly
***Medetera* sp.**
Larvae have an elongated, transluscent, shiny white body. The head is pointed, with black internal mouth parts visible, and the end of the cylindrical abdomen is bluntly rounded. The larva feeds on bark beetle larvae and can be the most abundant predator of these insects.

Insects Associated with Spruce Beetles (continued)

Dendrophagus cygnaei

Sylvanid beetle larvae and adults live under the bark and are likely predators of bark beetles. Commonly found associated with spruce beetle infestations.

Lyctocoris

This predator bug feeds on bark beetle larvae.

Flowerfly

The larval body is flattened, wedge-shaped, narrow at the front, and broad toward the back. The texture is leathery and not glossy. These are opportunistic, generalist predators.

Cecidomyiidae

Larvae are either gall makers, aphid predators or feed on fungi. They are often found in bark beetle galleries.

Between Two Worlds: Air and Water

The surface of water bodies makes an abrupt and distinct transition zone between two very different habitats. Small terrestrial insects accidentally blown over water often get stuck when they come in contact with the water's surface. At their small scale, the surface tension of the water is overwhelming and they get pinned down, unable to lift their wings from the surface. If they are not immediately gobbled up, they either sink or drift until they contact an emergent object. Shorelines provide a steady supply of helpless arthropods and are inhabited by a specialized guild of riparian scavenger-predators. These scavengers work their trade from both sides of the water surface. Hunting for disabled insects, small shore beetles, *Elaphrus* and *Bembidion*, scout the waterline, whereas striders search the water surface and diving beetles hunt underwater.

Amphibious species that spend part of their lives on each side of the surface have adapted behaviors to minimize their loss during the particularly vulnerable time of crossing between the two media: many insect adults fly off instantly from their pupal skin as soon as they bob to the surface; others crawl up the emergent vegetation at night. Water striders have their tarsi (feet) coated with a water repellent substance which allows them to walk on top of the water surface.

Elaphrus sp. ground beetles specialize in attacking terrestrial insects stranded in the water surface along the shoreline.

***Gyrinus* sp.**

Whirligig beetles spin at high speed on the water surface of eddies near the shore. They have 2 pairs of eyes, one facing above and one below water. They are common on lakes and slow rivers.

Aquatic Insects

Temporary Pools

Insects which live in temporary pools or ditches fed by seasonal spring fluctuations have short life cycles and the means to survive dry periods, as a drought resistant egg, larva or adult. These insects either remain inactive at the site until drought conditions improve or as active terrestrial adults. Temporary pools lack predatory fish and can therefore support rich invertebrate communities. In the late spring, at their peak production, such temporary pools are teaming with life.

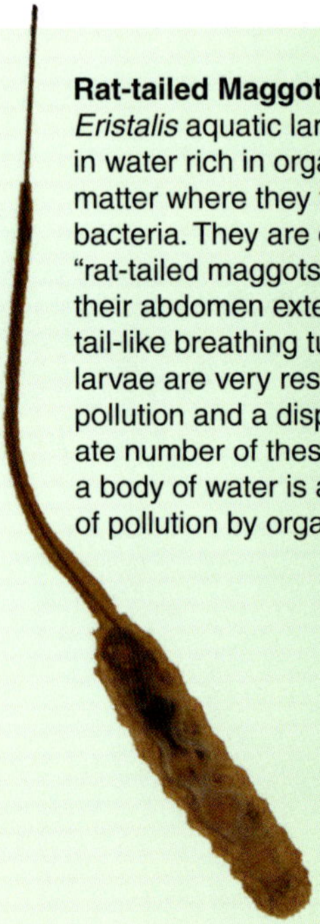

Rat-tailed Maggot
Eristalis aquatic larvae live in water rich in organic matter where they feed on bacteria. They are called "rat-tailed maggots" because their abdomen extends into a tail-like breathing tube. These larvae are very resistant to pollution and a disproportionate number of these larvae in a body of water is a sure sign of pollution by organic matter.

Mosquitoes
Larvae of the mosquito *Culiseta impatiens* develop most often in semi-permanent pools where the survival of the larvae is fairly high. However it is not uncommon for temporary pools to dry up prematurely, killing the larvae before they have a chance to mature. The larvae feed on plankton by filtering water. Movement of various mouthparts creates turbulence, directing large amounts of water through the mouth area.

At regular intervals, the mosquito larvae come to the surface and take in oxygen through the tube extending past their abdomen. They swim by wiggling from the head to the end of the abdomen.

Phantom Midge
Phantom midge larvae float horizontally, still in mid-water. Their body is almost transparent, hence their name. Their antennae are modified to grab minute prey. These strange-looking larvae can be found in temporary pools.

Swift Rivers

The water of fast-moving streams is well-oxygenated and carries nutrients in suspension within the water column. Insects living in these waters cling to the rocky river bottom or hide under stones and pebbles. If they are inadvertently caught in the fast current, they must swiftly sink back to the safety of the riverbed. Most of these bottom dwellers are passive food catchers, collecting drifting organic particles with various filtering devices. Well-oxygenated subalpine creeks appear to be especially rich in insect species and density.

Mayfly larvae
These larvae have a very low profile. All of their external structures, lateral gill plates and head cover plates are slightly tilted, pressing the larva against the rocks on the streambed. This mechanism can be demonstrated by holding a hand flat in the wind through the window of a moving car. When tilting the hand downward into the wind, the hand is forced downward.

Blackflies
Blackfly larvae attach to rocks or other stable substrate in flowing water with the aid of tiny hooks on the end of their abdomen. They have modified mouthparts, called labral fans, which enable them to filter and eat organic particles from the water.

Cranefly larvae
These predaceous larvae crawl in shallow water within the gravel bed avoiding exposure to the full force of the current.

***Brachycentrus* sp.**
Larvae of this caddisfly live only in cool running water. With silk, they attach the edge of the anterior opening of their case to a rock in the current, letting the water flow through the case. The larvae feed by extending their fore and middle legs to the side of the case and collect food particles in the hairs of their legs.

Perlodidae
Like most other families of the Plecopteran order, Perlodidae larvae live in cool, clean running water. Each species tends to be limited in the type of microhabitat it exploits.Each species has a marked preference for specific substrates such as cobbles, coarse gravel, leaf litter etc.

Aquatic Insects (continued)

Lakes

Many aquatic insects are found in vegetation that develops along the shoreline of sheltered lakes where there is little disturbance caused by waves or extreme fluctuation of the water level. This vegetation offers some protection from predatory fish.

Dragonfly
The larval mandibles are mounted on an articulated "arm" that projects forward to grab prey.

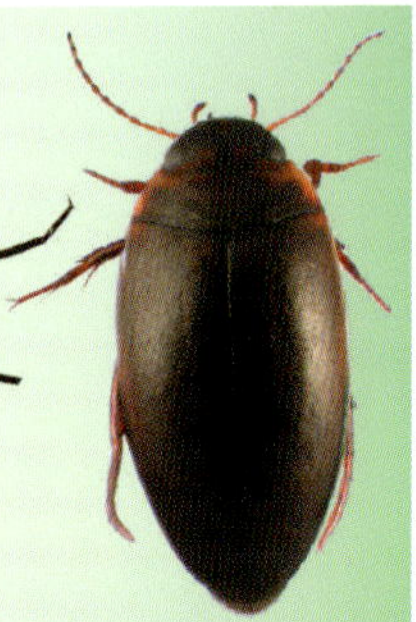

Diving Beetles
Both larvae and adults are aquatic predators. The larvae are armed with a pair of sickle-shaped hollow mandibles used to inject a digesting fluid into their prey. They suck back the liquefied nutrients, leaving behind an empty skin. Diving beetles attacks invertebrates, small fish and frogs.

***Plateumaris* sp.**
The larvae are aquatic, feeding on submerged stems of lake plants.

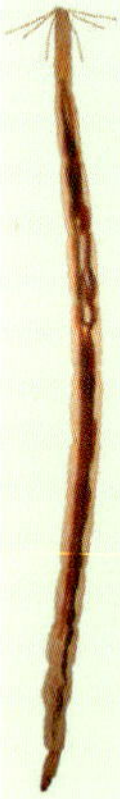

No-see-um
The larvae live in wetlands and in muddy bottoms of lakes. They can reach very high densities.

Caddisfly

Caddisfly larvae fabricate a casing with materials found in their habitat.

Aquatic living conditions vary greatly and most aquatic insect species have well-defined ecological needs and habits, limiting them to very specific microhabitats. This is clearly observed in the case of immature caddisflies. Each species of caddisfly favors very specific habitats which dictate the choice of materials used in the construction of their "case." Most genera are so consistent, with regard to material used, that their case alone is sufficient to insure a proper identification.

Household Pests

Carpenter ant damage in urethane foam insulation

Carpenter ants do not cause the decay of timbers nor do they eat the wood. They do, however, take advantage of any material easy to excavate such as urethane insulation or punky wood, in order to create their complex gallery network. Carpenter ants hunt small invertebrates tend aphids for honeydew and consume plant material. They get established, only when they have access to a reliable source of moisture.

Introduced, **cockroaches** are found in older buildings of large cities. Once established, they are difficult to completely eradicate.

Larder Beetles

Larder beetles are attracted by old meat, dog food and other meat products. During the warm season, they find their way into houses where they can flourish. The larvae are covered with long stiff hairs. Their telltale empty larval skins can often be found near their food source. Removing the food source will help eradicate the insect.

Pantry Grain Beetle
This minute introduced beetle feeds on spilled flour and other cereals. The beetles and their larvae multiply in flour bins or other wheat byproduct containers. The beetle can be eliminated by enclosing all wheat products in tight containers and by putting any food suspected of contamination into a freezer for a few days.

Spider Beetles
Spider beetles are some of the most common household pests in the Anchorage bowl. They are particularly attracted to dog food.

The Birch Leafminer, Sawflys, and spruce beetle

Insect outbreaks are a fact of life in south-central Alaska. Currently, the introduced birch leafminer, *Profenusa thomsoni* , defoliates birch trees over thousands of acres. The introduced larch sawfly, *Pristiphora erichsonii,* causes havoc on introduced Siberian larch in the Anchorage suburbs and on the Kenai Peninsula. The introduced woolley alder sawfly ***Eriocampa ovata,*** skeletonizes the leaves of entire alder stands, and for more than a decade now, the spruce beetle has been remodel the landscape of the western Kenai Peninsula and Cook Inlet area by killing most of the mature white spruce trees growing in this region. How natural are these phenomena? Insect outbreaks are not new. Through their studies of tree rings, Ed Berg, Ecologist with the U.S. Fish and Wildlife Service, and other scientists have shown evidence of cyclical spruce beetle outbreaks occurring twice a century, and there is evidence of bark beetle outbreaks as early as the 1760s on the Kenai Peninsula. However, recent outbreaks are longer in duration and subsequently more intense than those of the past, likely due to a serves of warm summers brought about by climate change.

Woolly alder sawfly larva. Some of the waxy secretion has been removed from this larva.

Eriocampa ovata

Alder sawfly outbreak near Kenai Lake, on the Kenai Peninsula. The leaves of the alders in the foreground are completely defoliated by sawfly larvae. All that remains of the leaves are a skeletons of the main veins. A few alder shrubs in the same thicket appear to be resistant and remained untouched.

Insect outbreaks are unusual events. In a stable natural community, over the long run, each individual organism is replaced by its offspring at an average ratio of 1:1. The equilibrium results from the interaction between the capacity of a species to generate offspring and the impact of combined non-living (abiotic) and biological (biotic) factors compromising the survival of those offspring before they reproduce. In other words, the equilibrium results from the brood size and the mortality incurred by the brood prior to reproduction. This mortality is due to a combination of adverse weather conditions, food limitation, and lethal organisms such as diseases, parasites, predators, and parasitoids. Parasitoids and specialist predators are more efficient than generalist predators at finding insects that live in concealed locations such as leaf miners and bark beetles. Outbreaks of introduced species can often be tracked to the lack of those parasitoids that keep the insect in check within its region of origin.

For spruce beetles, the generally accepted explanation for outbreaks is that spruce trees are drought stressed during early warm springs and lose their capacity to repel the spruce bark beetle by flooding their galleries with resin. Also, there has been an increase in the number of beetles that are able to complete their life cycles in one year rather than the normal two years.

Extensive forests dominated by a single tree species provide favorable conditions for the development of an outbreak. Climatic conditions also become paramount in the evolution of an outbreak. Harsh weather conditions, such as cold winters with little snow cover, cause high mortality rates in the many species of insects that overwinter in the duff layer.

It is important to maintain a diverse population of insect species because some predators or parasitoids, no matter how few, may have an important role in controlling the population of unfavorable insect populations.

Noctua pronuba

Photo Yves Dubuc

The European cutworm, *Noctua pronuba,* was accidentally introduced in Halifax, Nova Scotia in 1979. From there, it spread to Newfoundland in 1984, to Maine in 1985, and to Wisconsin in 1996. Butterfly specialist Ken Philip, of Fairbanks, Alaska, captured 27 of the moths in Haines, Alaska in 2005. This was the first time the species was found in Alaska. It is expected that the moth will soon reach the Matanuska-Susitna Valley. A generalist herbivore, *Noctua pronuba* is an agricultural pest. The females are especially fecund, each laying some 450 eggs in a season.

The moth is quite distinctive with yellow underwings and a broad black bar on the hind wings.

Massive Aphid Emergence

On June 1st, 2005, Erik Huebsch and Dr. Stephen Okkonen spotted what appeared to be an oil slick a few miles offshore of the mouth of the Kenai River in Cook Inlet. Dr. Okkonen, a University of Alaska Fairbanks physical oceanographer, was returning from a routine survey of the Inlet temperatures and salinities between the Forelands.

On close examination, the apparent slick turned out to be a 1/4" thick floating mat of insects that had collected in a tidal rip. Conservatively, Dr. Okkonen estimated the mass of insects to be about 2-3 feet wide and more than a mile long, which computes to a volume of between 8 and 12 cubic yards! Enough to fill a full-size dump truck!

The insects were winged aphids, identified by Dr. R. Foottit as *Boernerina occidentalis,* which lives on alders.

Early May was warm and sunny, so the aphids, which were tapping the abundant and nutritious spring sap, thrived and reproduced at a fast pace. By the end of May, winged aphids were numerous on the stems.

We speculate that the wind lifted up the winged aphids as they were in a synchronized dispersal phase and kept them suspended en masse in mid-air, turning the weak fliers into an aerial plankton. A large number of the aphids ended up being blown over Cook Inlet and eventually fell onto the water surface. Because few surface-feeding insectivores are present on the murky waters of Cook Inlet, most of the aphids were not eaten as they struggled on the water surface. Petrels, sea gulls and many sea ducks are surface skimmers, but their numbers are low and their impact must have been minimal facing the magnitude of this insect downpour.

Complex currents and winds skimmed the floating carcasses and concentrated them in the tidal rip. Such an astounding mass of carcasses of one type of insect in one location is quite extraordinary, and probably only a fraction of the local *Boernerina occidentalis* aphids that ended up in the Inlet.

Excerpt from Peninsula Clarion, June 24, 2005

The alder aphid *Boernerina occidentalis,* as they float en masse in a tidal rip of Cook Inlet. The striped coloration of the abdomen is typical for the species.

Part 2

Classification and Identification

Classification of Living Things

Carl Linneaus, born in 1707, initiated a system by which all living things are given a name composed of two words: the generic and the specific epithets. Over 1.7 million species have been named to date, of which insects number over a million. It is estimated that the total number of species of all living things is on the order of 10 million, with only 1% larger than a fly.

Linneaus' original classification system was improved to organize all living things according to their affinities such that species that share a common ancestor are grouped together. This classification system is still being fine-tuned today, as more information is gathered through DNA studies. Cladistics studies, in particular, focus on determining what ancestral traits are common between two organisms, thereby determining their degree of affinity. This is not always obvious as two similar-looking organisms might not be closely related (see p. 27). Changes in classification resulting from recent studies can be annoying when the new information causes major changes like the transfer of one species from one genus to another or the fusion of two families. Like anything else in science, nothing is written in stone and all is subject to revision. The classification used in this book reflects the latest changes.

The highest level of classification is the Kingdom: e.g., Monera (bacteria), Protista (protozoa, for example *Trypanosoma avium*, p. 44), Fungi (molds, mushrooms) Plantae (flowering plants, ferns, mosses), and Animalia (insects, mammals, birds etc.).

The Kingdom Animalia is divided into 16 Phyla (plural of Phylum) including Arthropoda.

Phylum Arthropoda

Contemporary non-marine arthropods found in south-central Alaska are divided into four subphyla: Crustacea, Myriapoda, Chelicerata and Hexapoda.

Subphylum Crustacea

ISOPODS
ORDER ISOPODA

Photo Tom Murray

Sowbug
Oniscus asellus
These introduced sowbugs are common terrestrial crustaceans found in house basements. They feed on decayed vegetation.

Subphylum Myriapoda

STONE CENTIPEDES
ORDER LITHOBIOMORPHA

Length: to 1". Adults have 15 pairs of legs. They are common under rotten logs and under stones and a have a flattened body.

Centipede

Subphylum Chelicerata

PSEUDOSCORPIONS
ORDER CHELONETHIDA

Halobisium occidentale
Length: 1/8". This minute pseudoscorpion is common in loose rocks of coastal beaches of south-central Alaska, at the high tide mark. It does not have a stinger and is not dangerous.

MITES
ORDER ACARI

***Poecilochirus* sp.**
Length: less than 1/32". Adult mites have 8 legs, immatures have 6. The head and the thorax are fused as in spiders. These minute arthropods are everywhere and have a variety of life strategies: parasite, predators, herbivores, detritivores.

SPIDERS
ORDER ARACHNIDA

Steatoda borealis
Length: 3/8". This spider spins cobwebs in sheltered recesses, capturing invertebrates that become entangled in the web. Common around houses and in rocky areas.

HARVESTMEN, DADDY LONGLEGS
ORDER PHALANGIDA

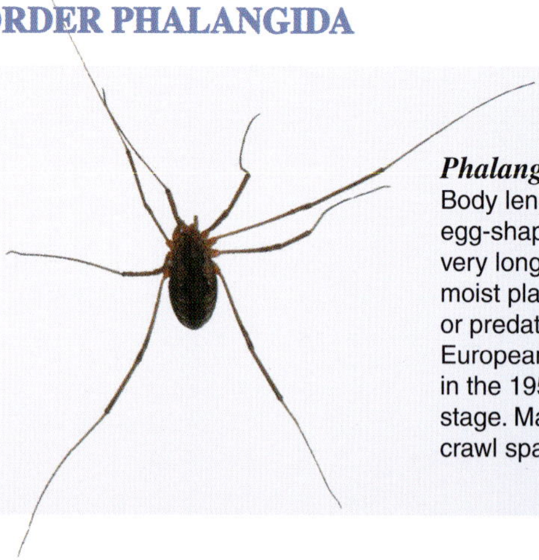

Phalangium opilio
Body length: up to 3/8". The body is egg-shaped, the legs are slender and very long. These animals are common in moist places. They are either scavengers or predators of small insects. This is a European species, first noticed in Alaska in the 1950s. They overwinter in the egg stage. Masses of adults overwinter in crawl spaces.

Subphylum Hexapoda

The subphylum Hexapoda (six-legged) includes the classes Entognatha and Insecta.

Class Entognatha

The Class Entognatha comprises the orders Diplura, Protura and Collembola. All are minute insects. The order Collembola is presented here.

SPRINGTAILS
ORDER COLLEMBOLA

These primitive arthropods are very common but are often overlooked because of their minute size. They do not have wings and their eyes are simple. Collembola live in the leaf litter in great numbers.

To find Collembola, deposit a fistful of fertile soil from a flower pot or hummus from the garden on a white plate. Springtails will stand out against the white background, jumping erratically.

Family Poduridae

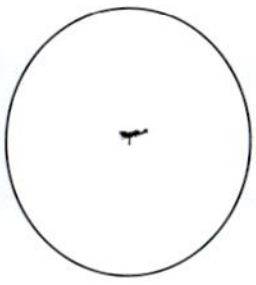

Length: less than 1/16". Their body is hairless. Podurids live deep in the soil.

Family Entomobryidae

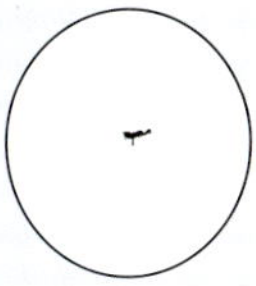

Length: less than 1/16". The springtail's body is covered with fine hairs or by a wax coating. Their abdomen extends into a fork that is folded under, spring-loaded. They quickly release the forked tail to propel themselves in the air thus their common name "Springtails". They live on the upper layers of the leaf litter.

Globular Collembola
Family Sminthuridae

Photo by Matt Bowser

Ptenothrix sp.
Length: less than 1/32". In profile, they resemble a sitting rabbit. Globular collembola are common in grassy areas.

Insect Classification

Class Insecta

Insect species have evolved from a single species over many millions of years. Descendants of the original primitive population differentiated into an array of variants. The scientific classification system attempts to reconstruct this evolutionary process.

Example	Carpenter ant
Kingdom	**ANIMALIA**
Phylum	**Arthopoda**
Subphylum	**Hexapoda**
Class	**Insecta**
Order	**HYMENOPTERA**
Family	**Formicidae**
Subfamily	**Formicinae**
Genus	***Camponotus***
Species	***Camponotus herculeanus***

In this book, we will be referring mostly to the last 5 levels: orders, families, subfamilies, genera (plural of "genus") and species.

By convention, the order name ends with "a", the family name ends in "-idae", the subfamily in "-inae", the first letter of the genus is capitalized and the specific epithet is written in lower case letter.

Carpenter ant queen

Scientific Names

Common names of animals vary among languages and even regionally within the same language. This can be misleading. For instance, aphids are sometimes called "plant lice" even though they are not closely related to lice. To remedy this confusion, scientists have adopted a universal system (*) in which each species is given a unique name. This name is composed of two words always written in italics. The first word, the genus name, is followed by the specific epithet. The pair of words are unique for each animal species. For instance, the scientific name of the domestic cat is *Felis catus*, the cougar is *Felis concolor,* the tiger is *Felis tigris*. Scientific names are coined as need arises with the publication of the description of new species. The names can be descriptive, refer to people or places and are "Latinized."

When an animal is identified only to the genus level, the species name is replaced by "sp." meaning "species unspecified." Example: *Platygaster* sp.

The pronunciation of scientific names is not as consistent as its spelling. One could even say that scientific names are almost like Chinese script: every Chinese can read the written word, but inhabitants of two distant regions in China may be unable to communicate verbally. Every biologist can read the Latinized scientific name, but two biologists of different language background might not recognize at first a name when pronounced by the other. Different authorities have tried to standardize the pronunciation based on historical interpretation of the Latin pronunciation.

(*) Almost universal would be a more correct statement: in a few rare cases, a plant and an animal share the same name. For instance, the genus name of the louse *Pedicularis* is also the genus name of the plant lousewort.

Identification key

To use an identification key, read and compare the first pair of statements, 1.a and 1.b. Choose which of the two statements best applies to the insect you want to identify. At the end of the chosen statement, the number indicates the pair of statements to progress to in the identification. Resolve that next pair of statements. Follow the same procedure until you end up with the order.

Key to the Orders of Adult Insects Included in this Book

1.a 3 pairs of legs and body with 3 distinct parts, head, thorax and abdomen, sometimes wings. Most adult insects..**2**

1.b Not 3 pairs of legs, body not in 3 distinct parts and no wing........................ **Not an adult insect**

2.a One pair of wings **3**

2.b Two pairs of wings or no wing **4**

3.a One pair of forewings. Hindwings reduced to a club shaped stub (haltere).................................. **Flies, midges, mosquitoes: Diptera. See p. 128**

3.b Forewings reduced to a stub, hindwings functional, fan-like, small, Rare insect................. .... **Strepsiptera. See p. 127**

4.a Two pairs of wings **5**

4.b No wings..**19**

5.a Wings narrow, long, bordered with a fringe of long hairs, insect less than 1/32" long, body elongate..........**Thrips: Thysanoptera. See p. 93**

5.b Wings broader, hairs, insect size and shape variable...**6**

6.a Wings large, covered with a fine powder (scales), mouth parts a coiled tube................................... ....**Butterflies, moths: Lepidoptera. See p. 143**

6.b Wing not covered with fine powder (scales), mouthpart not a coiled tube...............................**7**

7.a Mouthparts a stiff piercing tube **8**

7.b Mouthparts not a stiff piercing tube **9**

8.a Forewings partially hardened, opaque................... .... **True bugs: Heteroptera. See p. 87**

8.b Entire wings transparent except the veins............ ......**Aphids, psylids leafhoppers: Homoptera.** **.. See p. 90**

Diptera

Strepsiptera

Thysanoptera

Lepidoptera

Hemiptera

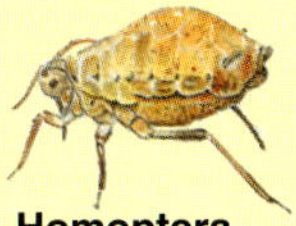

Homoptera

Orthoptera

Coleoptera

Dermaptera

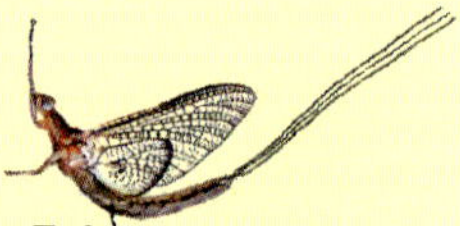

Ephemeroptera

Hymenoptera

Plecoptera

Blattoptera

9.a Hind legs strongly developed for jumping, wings stiff, 3/8" and longer.. **Grasshoppers, crickets: Orthoptera. See p. 84**

9.b Hind legs not strongly developed for jumping or whole body shorter than 3/8", wings variable..**10**

10.a Forewings hardened, often opaque, forming a shell protecting functional hindwings...............**11**

10.b Wings transparent or translucent except for the veins..**12**

11.a Hardened forewings covering most or all the abdomen, or if covering only the anterior portion of the abdomen, no pincer-like appendices at the end of the abdomen..**Beetles: Coleoptera. See p. 102**

11.b Hardened forewings covering only the base of the abdomen, conspicuous "pincers" at the end of the abdomen.. **Earwigs: Dermaptera. See p. 85**

12a Forewings much larger than hindwings...........**13**

12b Forewings and hindwings almost the same size..**14**

13.a Antennae as long as or longer than the body, 2 or 3 long filaments extending beyond the end of the abdomen, silhouette typical, wings erect and back to back when at rest....................**Mayflies: Ephemeroptera. See p. 96**

13.b Antennae shorter than body length, never 2 or 3 filaments extending beyond the end of the abdomen, sometimes 1 long stiff ovipositor, wing position at rest variable........................**Wasps, bees ants, sawfling: Hymenoptera. See p. 160**

14.a Body flattened, soft, wings laying flat on the body when at rest..**15**

14.b Body not flattened, no obvious filament extending beyond the abdomen **16**

15.a Body elongated, two long filaments extending beyond the abdomen and the wings, found near rivers with swift current........................**Stoneflies: Plecoptera. See p. 98**

15.b Body oval, filaments at the end of the abdomen, short if present, found inside dwellings................**Cockroaches: Blattoptera. See p. 85**

16a Wings held roof-like when at rest, abdomen variable, eyes not globose, body not extending beyond the wings.. **17**

16.b Wings held spreading or gathered above the abdomen, abdomen long extending beyond the wings, eyes very large, globose........................... .... **Dragonflies: Odonata. See p. 94**

17.a Body less than 3/16" long.................................... .... **Psocoptera. See p. 83**

17.b Body more than 1/4" long.............................. **18**

18.a Wings transparent, hairless with “lacework-like” network of veins, antennae equal or less than half the length of the body,................................. .... **Lacewings: Neuroptera. See p. 86**

18.b Wings more or less opaque, hairy toward the front, antennae as long or longer than the body... .... **Caddisflies: Trichoptera. See p. 99**

19.a Parasitic on mammals or birds **20**

19.b Not parasitic on mammals or birds................. **21**

20.a Body flattened laterally, long legs, jumping insect..............**Fleas: Siphonaptera. See p. 127**

20.b Body flattened vertically, crawling insect **22**

21.a Body pear-shaped, longer than wide .... **Louse: Phthiraptera. See p. 83**

21.b Body as wide as long... .... **Crab louse: Phthiraptera. See p. 83**

22.a Ant-like insect.. .... **Wingless Hymenoptera. See p. 160**

22.b Not ant-like **23**

23.a Body elongate, covered with scales, long antennae **24**

23.b Not as above .... **Collembola. See p. 77**

24.a Eyes large, touching in the middle, 2 short lateral and one long central filaments........................... .... **Bristletails: Archeognatha. See p. 82**

24.b Eyes small, not touching in the middle, 3 long filament at the end of the abdomen, the laterals divergin.......... **Fire Brats: Thysanura. See p. 82**

Odonata

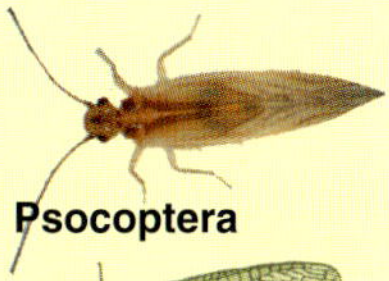
Psocoptera

Neuroptera

Trichoptera

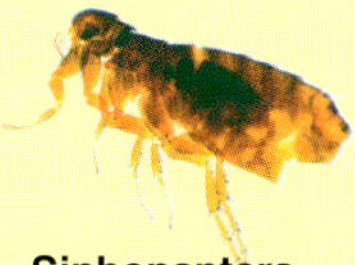
Siphonaptera

Anoplura

Phthiraptera

Collembola

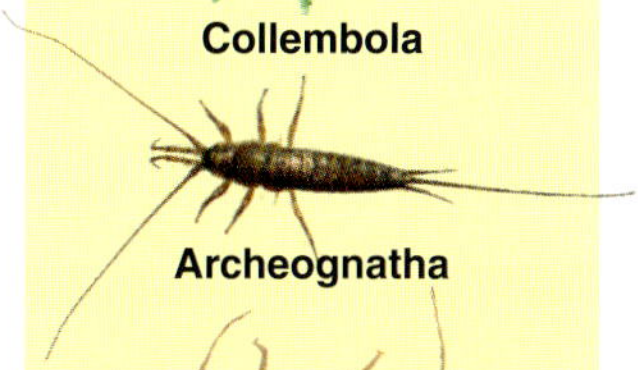
Archeognatha

Thysanura

JUMPING BRISTLETAILS
ORDER ARCHEOGNATHA
Family Machilidae

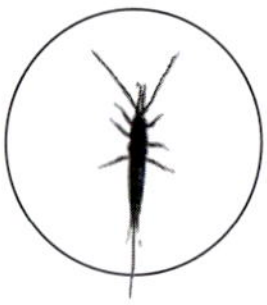

Petridiobius arcticus
Body length: 3/8". Bristletail eyes meet at the top of the head, and the two lateral bristles at the end of the abdomen are short and point backward. *Petridiobius arcticus* live in loose colonies amongst the rocks within the high tide zone where they feed on lichens. They jump to escape predators. To find them, carefully roll over rocks along the shoreline. Individual bristletails may be seen basking in the sun but otherwise they are mostly nocturnal. *Petridiobius arcticus* is known only from southern Alaska.

BRISTLETAILS
ORDER THYSANURA
Family Lepismatidae

Firebrats
Thermobia domestica
Length: 3/8". Similar to other bristletails, but the 2 lateral bristles at the end of the abdomen are long and point outward. The eyes are widely separated on top. Adults live up to 2 or 3 years and may molt up to 50 times. These primitive insects are found in food pantries where they feed on starchy goods. They are nocturnal and in Alaska survive only in heated buildings.

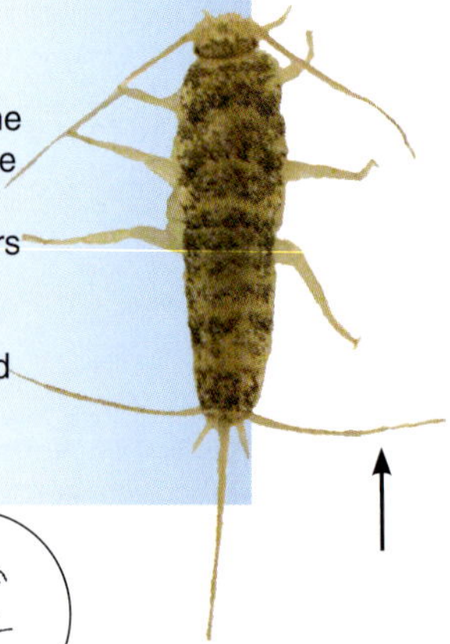

BOOK LICE
ORDER PSOCOPTERA

Common Book Lice, Psocids
Family Psocidae

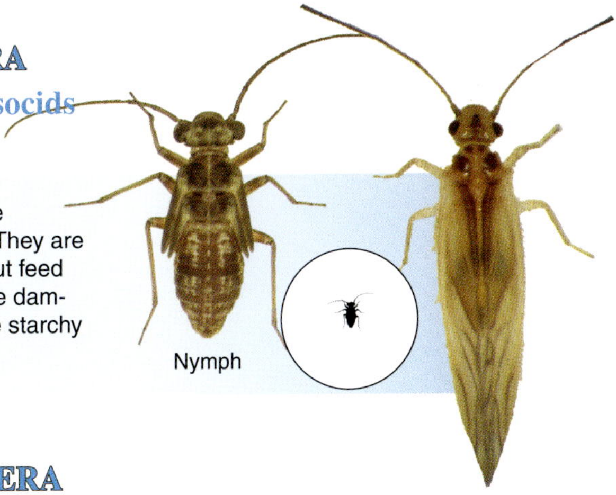

Length: to 1/8". The wings are folded roof-like when at rest. They are not related to parasitic lice, but feed on dry goods. They can cause damage to books when eating the starchy glue used in older bindings.

LICE
ORDER PHTHIRAPTERA

These insects are parasitic on mammals. Most are specialists and each species is associated with a single mammal species. Two species of these lice (singular: louse) have been parasitic on humans since the early years of humanity. They are the human louse and the crab louse.

Louse
Family Pediculidae

Human Louse
Pedicularis humanus
Length: to 1/8". The body is flat, pear-shaped and without wings. This is the head louse that is chronically spotted by school nurses. The entire life of this louse is spent on humans. The eggs, called nits, are attached to the hair and immature nymphs resemble a smaller version of the adult. The whole life cycle lasts about a month. Lice are transmitted from one person to another through common use of a comb, clothing or bedding. Lice are the vector of typhus. Lice infestation is treated with a combination of fine tooth combing and application of specific insecticides.

Crab Louse
Family Pthiridae

Pthirus pubis
Length: to 1/8". The crab louse is broader and shorter than the head or body louse. This species prefers remaining in the pubic hair. Occasionally they are also found in horse eyelashes. Crab lice die if separated from the human body for more than twenty-four hours. The eggs, which are laid in clothing, take one week to hatch.

GRASSHOPPERS AND CRICKETS
ORDER ORTHOPTERA

Grasshoppers
Family Acrididae

Grasshoppers feed on grasses, and occasionally on dying invertebrates. They often bask in the sun on exposed bare mineral soil. Migratory grasshoppers and other species cause extensive damage to barley, wheat and vegetables in the Lower 48 states and in Delta Junction, Alaska, where they have been known to migrate in dense swarms.

Migratory Grasshopper
***Melanoplus* spp.**
Length: up to 1". The thorax does not extend over the abdomen. These grasshoppers are found in cool damp locations, near ponds, in forests, on grassy mountain slopes, along roads and above treeline. They browse on grasses and forbs. In Fairbanks, Alaska, the eggs hatch only after having been submerged in melting snow water. The nymphs remain near the location where the eggs were deposited. The winged adults disperse.

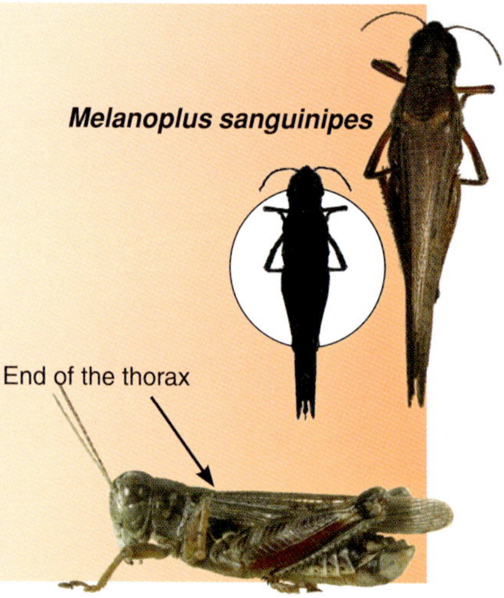

Pygmy Grasshoppers
Family Tetrigidae

Tetrix ornata
Length: 1/2". The thorax of the pygmy grasshopper extends to the end of the abdomen. They are found on poorly vegetated soils near water. Some species are good swimmers.

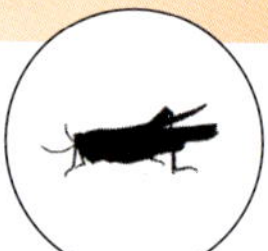

Crickets
Family Gryllidae

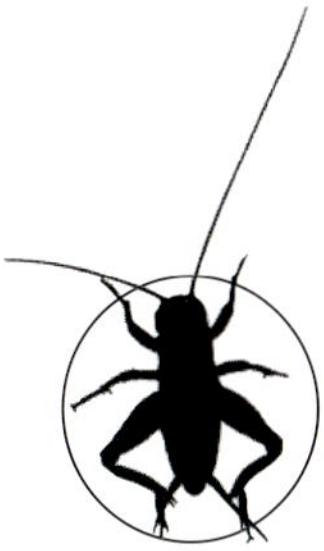

Domestic Cricket
Acheta domesticus
Length: 5/8". This is the domestic cricket sold in pet stores. It occurs naturally in deserts of North Africa and southwest Asia. This introduced species is well-adapted to living without water. Escapees will survive on their own, hiding behind sheetrock panels during the day and venture out at night to feed. If they have access to food, they may reproduce. This cricket emits a stridulation that can be heard occasionally in public buildings.

COCKROACHES
ORDER BLATTODEA
Family Blattellidae

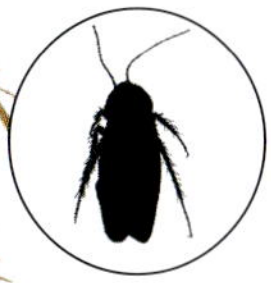

German Cockroach
Blattella germanica
Length: 1/2". Originally from Africa, this species came to this continent via Europe, likely aboard some of the first sailing ships. Cockroaches are nocturnal. German cockroaches can be found in some of the older buildings of the older cities of south-central Alaska. Cockroaches are not known to carry diseases.

EARWIGS
ORDER DERMAPTERA
Earwigs
Family Forficulidae

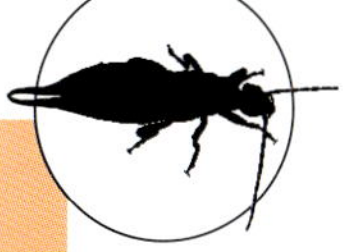

Forficula auricularia
Length: 3/4". This introduced species has been found in houses in Anchorage. It is not likely to survive the winter outside of heated buildings. Earwigs are predators of smaller invertebrates. Their name comes from the superstition that earwigs enter the ears of sleeping people.

NERVED-WINGED INSECTS, LACEWINGS ORDER NEUROPTERA

These insects have two pairs of broad and long membranous wings. The veins are numerous and often run parallel or perpendicular to each others. The mouth parts are of the chewing type.

Common Lacewings Family Chrysopidae

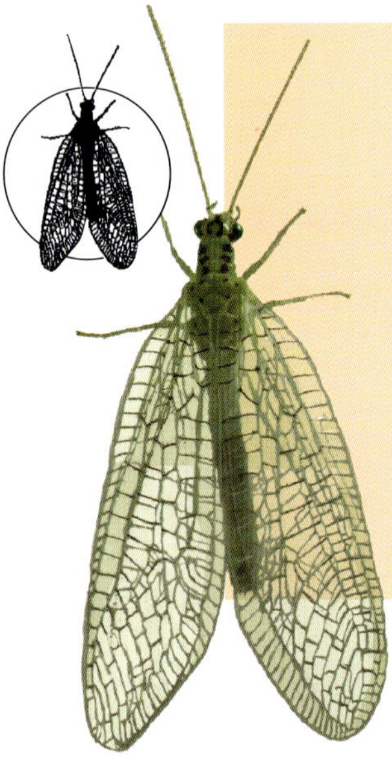

Common green lacewing head, side view showing the golden eye

***Chrysopa* sp.**
Length: 3/4". The long transparent, green-veined wings and the golden eyes are unique to this family. When at rest, the wings are held "roof-like" on the back of the insect. Both adults and larvae are voracious predators of aphids. They are often used in the biocontrol of aphids and can be purchased from greenhouse supply stores.

Brown Lacewings Family Hemerobiidae

***Boryomyia* sp.**
Length: 3/8". These are a dark brown, smaller version of the common green lacewing. Both adults and larvae are predaceous on soft-bodied insects like aphids.

ORDER HEMIPTERA

Suborder Heteroptera: True Bugs

In entomology, the use of the term "bug" is limited to insects of this order. One of the main characteristics of Heteroptera is that their mouthparts are modified into a piercing-sucking stylet. Eighteen families with one hundred fifty-eight species of Heteroptera have been identified in Alaska.

Shield Bugs
Family Acanthosomatidae

Birch Shield Bug
Elasmostethus interstinctus
Length: 3/8". This species feeds on the sap of many plants including its namesake, birch. The birch shield bug overwinters as an adult, mates in the spring and lays eggs in June on the upper side of the leaves and on birch catkins. The nymphs first remain grouped, then later disperse. This species feeds on birch, aspen, and devil's club and is often collected on willow, but may not feed on the shrub. This brightly colored insect is often noticed in the fall when they start entering houses in search of overwintering sites.

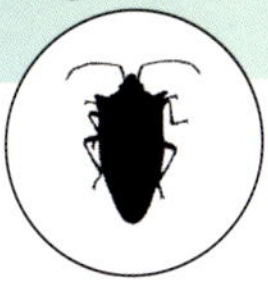

Flower Bugs
Family Anthocoridae

Length: to 3/16". The body is flattened above, black with white or red markings. These bugs are mostly generalist predators, opportunistic in their choice of prey. They are usually found in flowers or hiding under loose bark. Some species have been successfully used in biological control. Thirty species have been introduced accidentally in the US. Thirty three species and three genera are known in south-central Alaska.

This bug nymph feeds on bark beetle larvae.

Flat Bugs
Family Aradidae

There are seven species and two genera known in Alaska.

***Aradus* sp.**
Length: 3/16" to 1/2". These insects feed mostly on fungi. Five species are known in Alaska.

Flat in profile

***Aneurus* sp.**
Length: to 3/16". These extremely flat bugs can be found under the loose bark of decaying trees where they feed on fungi. They are especially abundant in fire-killed trees. Two species are known in Alaska.

Shore Bugs
Family Saldidae

This family has eleven species in six genera known from Alaska.

Length: 1/8". These small black bugs with globular eyes are common in sparse coastal vegetation and on mudflats at the water edge near the beach rye belt. They are often found in the same habitat as the tiger beetle. Shore bugs are predatory and run fast after other insects. When disturbed, they run or fly short distances.

Water Striders
Family Gerridae

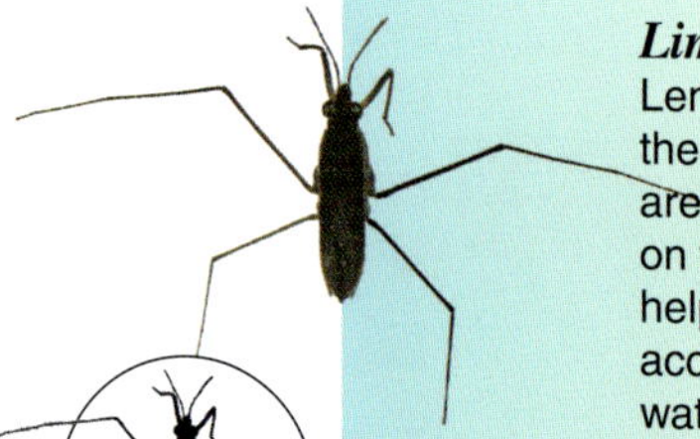

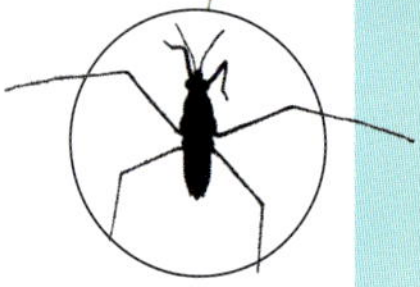

***Limnoporus* sp.**
Length: 1/2". The longer middle legs are used by the insects to propel themselves forward. Their feet are coated with a water repellent and act as skates on the water surface. Water striders feed mostly on helpless terrestrial insects that land in the water by accident. For their own protection from predators, water striders hide under bank vegetation. *Limnoporus* is one of two genera of water striders that can be observed gliding on the surface of the water at the margin of lakes and ponds and in eddies of rivers.

Water Boatmen
Family Corixidae

In Alaska, this family includes nine species in three genera.

***Hesperocorixa* sp.**
Length: 3/8". Boatmen, so called because the insects use the middle pair of legs like oars of a boat, are common in ponds and lakes. They feed on algae and microorganisms. Boatmen breathe air and come at regular intervals to the surface to replenish their supply, which they store in a bubble on the under part of their bodies. Excellent fliers, *Hesperocorixa* spp. exit the water in mid-August and fly great distances at night in search of new bodies of water to colonize. In a study in San Francisco, 72% of the prey (1% of the volume) found in the stomachs of juvenile king salmon were water boatmen.

Plant Bugs
Family Miridae

Length: 1/4" to 3/8". These bugs are common, especially on shrubby or grassy vegetation. They are mostly of dull coloration, but a few are brightly colored. Most of the plant bugs feed on plant sap and can be a nuisance when abundant. Others are predators and are beneficial to gardeners. There are seventy-three species in thirty-five genera known for Alaska.

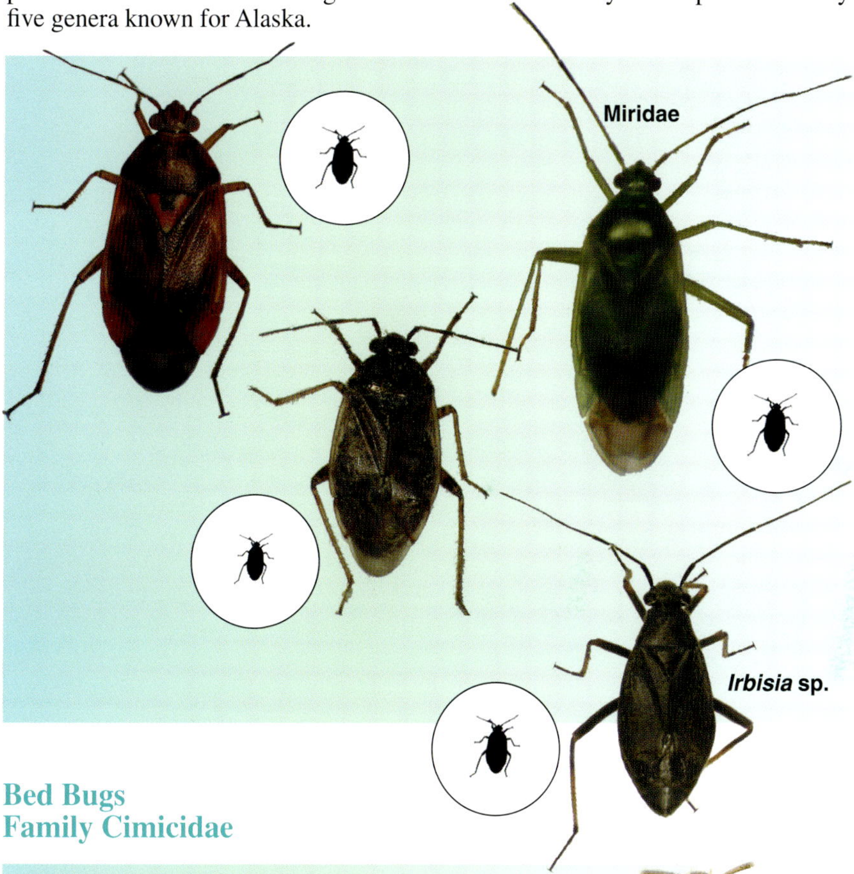

Bed Bugs
Family Cimicidae

Swallow Bug
Oeciacus vicarius

Length: to 1/4". This flat insect closely resembles the bed bug, but is associated with swallows nesting in bird houses. Where the bird houses are hung on the outside walls of inhabited houses, the insect might crawl inside the building through cracks. They occasionally bite people, but will not survive long away from their main hosts. The real bed bug *Cimex lectularius* bites humans but may survive on dogs, cats or even poultry. The eggs are deposited in cracks of bedding where the adults hide during the day. These bugs may be pests in high traffic hotels, and are known to be from every region of the state.

Suborder Auchenorrhyncha

The Auchenorrhyncha were formerly known as Homoptera. Most, if not all Homoptera, feed on plant sap. Like the Hemiptera, their mouthparts are highly modified into a piercing and sucking stylet. Species in some families like the aphids and the whiteflies are notorious pests.

Leafhoppers
Family Cicadellidae

The well-developed tibiae of these jumping insects are armed with two rows of fine spines. Some leafhoppers are specialists and feed exclusively on the leaves of a single or a few closely related plant species. They suck the sap by tapping into the phloem. Cicadellidae are attacked by the parasitoid dryinid wasps and big-headed flies. Leafhoppers can cause damage to their food plants by introducing pathogenic viruses or other microorganisms, or by damaging the structure of the leaf itself, causing the leaf to dry up prematurely. Some leafhoppers excrete honeydew like aphids. Eighty species distributed in thirty-seven genera are known for Alaska.

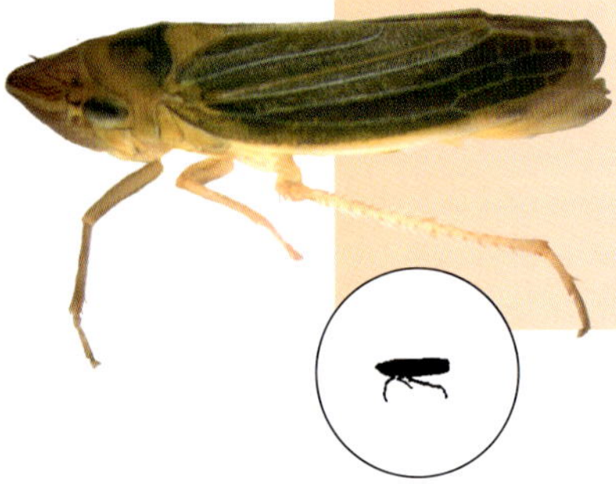

Draeculacephala angulifera
Length: 1/4". This bright green insect is one of our largest leafhoppers. Its Latin name translates as "Dracula's head". One of two species of the genus is found in south-central Alaska. It is fairly common in damp sites where it feeds on grasses and sedges.

Thamnotethix confinis
Length: 1/8". This smaller species is common in moist meadows. It is sometimes parasitized by dryinid wasps.

Jumping Plant Lice, Psylids
Family Psylidae

Psylids are reminiscent of cicadas by their shape, but their size is much reduced, rarely exceeding 1/4". Like leafhoppers, each species tends to feed only on a few closely related plant species.

Thirty-eight species in four genera have been identified in Alaska. An encyrtidae wasp has been found to parasitize psylid nymphs in south-central Alaska.

Cacopsylla alaskensis

Length: 3/16". This species lives on willows and is very common in south-central Alaska, especially on diamond-leaf willow. The adults overwinter and are one of the first insects to be active in the spring.

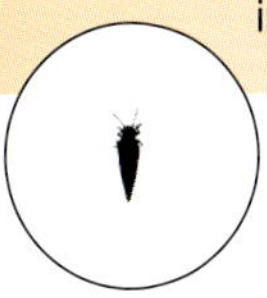

Alder Psylid
Psyla floccosa

The nymphs of this psyllid secrete filaments of wax that remain attached to the insect and protect it from enemies and inclement weather. The cotton-like masses are very obvious on alder in early summer.

Nymph with wax removed

Suborder Sternorrhyncha

Family Aphididae: Aphids

Aphids are quite distinct with their soft bodies and the two tubes (cornicles) located on their abdomen. They have complex life histories and switch plant species in the middle of the summer. Aphids are notorious agricultural pests. The density of a single colonies varies extremely in the course of a season and at the peak population density, aphids can cause extensive damage to plants.

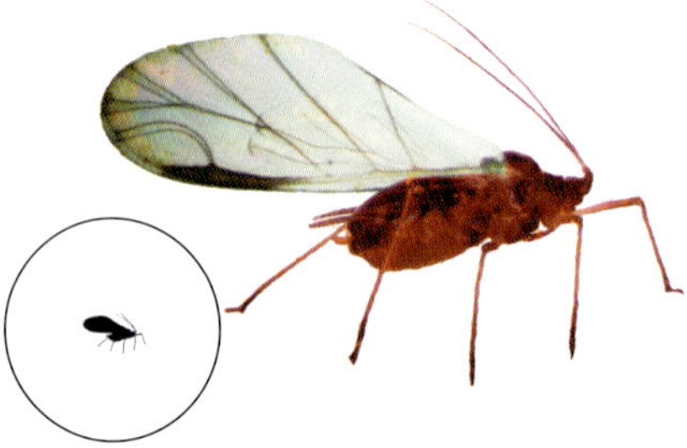

Black Willow Aphid
***Pterocomma* sp.**
Nymphs of these species are found on willow and cottonwood.

Alder Aphid
Boernerina occidentalis
Length: 1/16". The abdomen is yellow with transversal black bars. This aphid species is very common on alders in south-central Alaska. At the end of May or in early June, winged adults take flight and are wind dispersed.

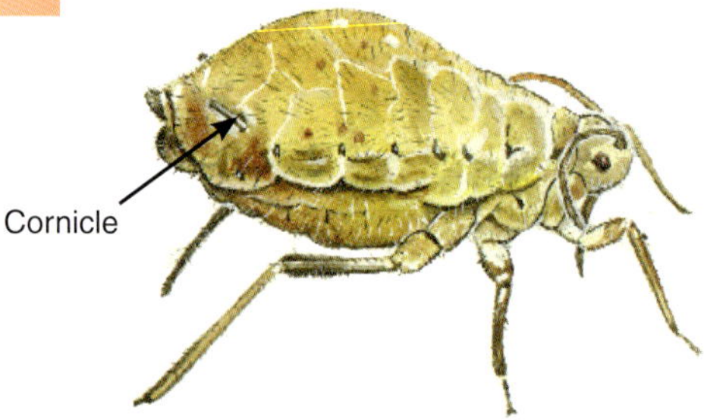

Aphid fundatrix. The orange eyes of the fully developed embryos are visible through the cuticule .

Spruce Gall Aphid
***Adelges* sp. and *Pineus* sp.**
Length: 1/32". These aphids cause stunting of new shoots, inducing the formation of cone-like galls up to 2" long at the tip of the branches. The nymphs living inside the galls are covered with a white powder and those living outside the galls are covered with a thick coat of waxy filaments like the alder psyllid nymph shown on page 91.

Spruce Aphids
Elatobium abietinum
Length: 1/32". The spruce aphid usually occurs on Sitka spruce. They feed on the needles and can reach high density causing defoliation during outbreaks that sometimes follow mild winters. This small aphid is capable of reproducing all year. They feed on old needles and when in high numbers can cause the death of the tree by causing premature drop of old needles, leaving only the young needles at the tips of the branches.

Giant Conifer Aphid
***Cinara* sp.**
Length: to 3/16". The giant conifer aphid feeds on branches and twigs of white and Sitka spruce trees causing yellowing of the needles but not massive defoliation. It is often found with spruce aphids.

THRIPS
ORDER THYSANOPTERA

Metamorphosis is intermediate between complete and incomplete: early stages are larvae, later stages are nymphs, pupa, then finally the fully-winged adult.

Common Thrips
Family Thripidae

Length: less than 1/16". When present, the 2 pairs of wings are long and narrow, fringed laterally with long stiff hairs. The mouth parts are of the piercing-sucking type and they either feed on plant juices or are predators. Thrips can complete a life cycle in 45 days. They can be a nuisance when numerous, causing the wilting of flowers and leaves. Predatory mites and ladybugs can be purchased from greenhouse supply outlets to control thrips in greenhouses. Garlic is reported to repel thrips which sometimes bite humans.

DRAGONFLIES
ORDER ODONATA

The lower mandibles of dragonfly larvae are articulated and can be projected forward (see p. 68). Dragonfly larvae are aquatic and can be found crawling on lake bottoms, in the shallows near the shoreline or hidden in the aquatic vegetation. The larvae surge forward by squirting water from the end of their abdomens. Fourteen species grouped in seven genera are known from south-central Alaska.

Skimmers
Family Libellulidae

Adult skimmers perch on a branch to stalk their prey. Once a flying prey is spotted, the skimmer quickly pounces on it in mid air and brings it to a perch to eat.

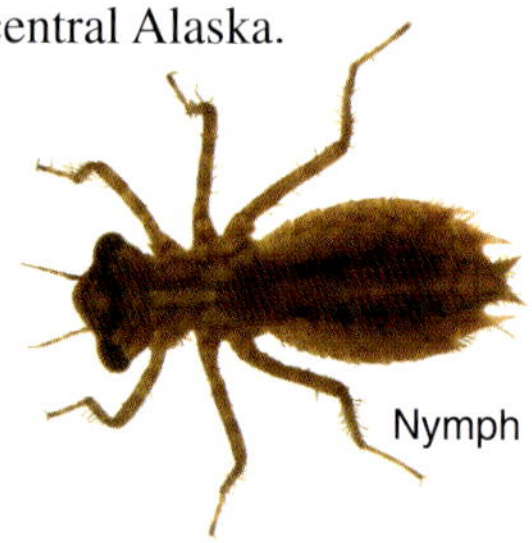
Nymph

1.3X

Male

Canada Whiteface
Leucorrhinia patricia
Length: 1". Wing span: 2". The eyes meet on top of the head. Males are entirely black. Females have a yellow pattern on top of the back, and the forehead is white for both males and females. They are commonly found on floating sphagnum moss bogs at the margin of small lakes.

Four-spotted Skimmer
Libellula quadrimaculata
Length: 1-3/4". The body is yellow-brown in the front, turning black toward the end of the tail. The leading edge of the forewings is banded yellow with a black spot in the middle. The base of the hindwings has a black triangular marking. This species lives in ponds surrounded by marshes. This dragonfly appears earlier than others in the spring and is widespread throughout Northern hemisphere. It is Alaska's state insect.

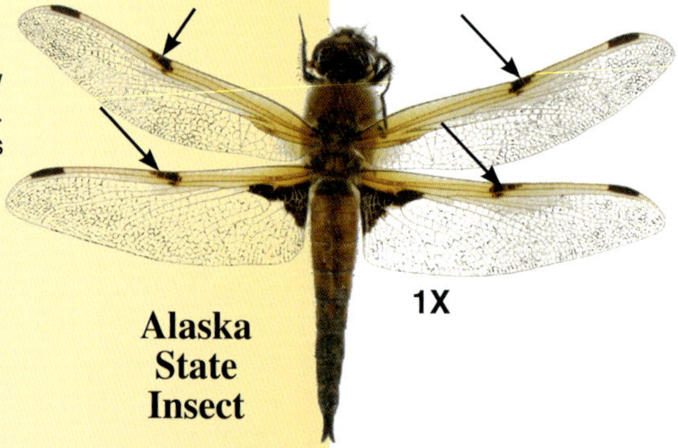
1X

Alaska State Insect

Pond Damsels
Family Coenagridae

These delicate damselflies rest with their wings gathered on their back. The nymphs of damselflies can be identified by the three feather-like extensions at the end of the abdomen. These small dragonflies are common on floating sphagnum mats located at the margin of small lakes. Two genera and three species are known to live within south-central Alaska.

1X

Boreal bluet
Enallagma boreale
Length: 1-1/2". There are two species of this genus in south-central Alaska. Theses insects are common in marshes and lake margins bordered by mosses.

Bluet at rest

Blue Darners
Family Aeschnidae

These powerful fliers can be found far from lakes where they were reared. The nymphs may take 5 years to develop to the adult stage. It is not uncommon to find the empty nymph skins hanging from the vegetation at the margin of lakes in midsummer. This family is well represented by six large species and includes the largest insects found in south-central Alaska.

Empty skin shed by emerging dragonfly

1X

Paddle-tailed darner
Aeshna palmata
Length: 2-1/4".

Mature larva

MAYFLIES
ORDER EPHEMEROPTERA

Mayflies are the oldest winged insects surviving unchanged from the Permian and Carboniferous geological periods. Imagoes (flying adults) are fragile, weak fliers and have soft bodies. Their long delicate wings are folded upward at rest and 2 or 3 long filaments, as long or longer than the body, extend beyond the abdomen. Imagoes do not have functional mouthparts and are unable to feed. Males can be distinguished from females by their larger eyes and pincers at the end of the abdomen. The adult stage is ephemeral (short-lived), a characteristic to which their scientific name refers, and lasts from a couple of hours to a couple of weeks. Mayfly nymphs live on fast current river bottoms or on the vegetation of freshwater streams and lakes where they feed on microscopic algae and on minute organic particles. Nymphs get their oxygen through a series of lateral gills visible on either side of the abdomen. Mayflies are unique among all insects for molting once more after they acquire fully functional wings. Mayflies are very sensitive to pollution. One species, *Palingenia* sp., has gone extinct in modern times from the Netherlands because of pollution in its habitat. Mayflies are an important source of food for salmon. Many fishermen use artificial flies patterned after emerging mayfly nymphs.

Six families and seventeen genera are known for south-central Alaska.

Family Heptageniidae

Epeorus longimanus

Length: 1/2". The most common mayfly in western America. The nymphs live in the upper sections of mountain creeks. The adults disperse in June or July. The under part of the nymph body is concave, forming a suction cup which helps the nymph adhere to rocks against the fast current. Nymphs are collector-scrapers, feeding on algae.

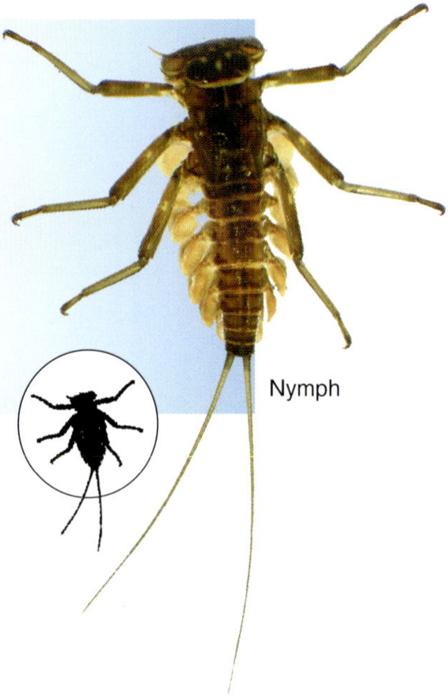

Nymph

Family Baetidae

Nymph

***Baetis* sp.**
Nymph body is less than 1/2" long. Dark colored with 2 or 3 filaments extending beyond the end of the abdomen. Adult females are brown while males are lighter. Hindwings are small or lacking altogether. The nymph body is cylindrical as species of this genus prefer quieter water, where dense vegetation grows. The nymphs swim freely in quiet waters and cling to the substrate where the current is stronger. Species of this widespread genus feed on diatoms collected by scraping the rocks.

Nymph

Family Ephemerellidae

***Drunella* sp.**
Nymph body length is 1/2". The body is strongly flattened, adapted for clinging to rocks in fast current. They feed by scraping rocks or by preying on other organisms. The genus is widespread.

Nymph

STONEFLIES: ORDER PLECOPTERA

Adult stoneflies keep their wings folded flat on their backs when at rest. Their antennae are at least half as long as their soft body. Adult stoneflies live 1 to 4 weeks, are mostly nocturnal, and hide during the day under stones or in vegetation. The aquatic nymphs are flat, have two filaments at the end of their abdomen and live in rivers with fast current or in cold lakes at higher elevations.

Some stonefly nymphs feed on salmon carcasses in the river and are themselves eaten by salmon fry. Studies in a Wyoming river showed that stoneflies constituted 10% of the diet of brook trout. Stonefly nymphs are very sensitive to the degradation of their habitat because they require clean, cool, well oxygenated water to live. Their presence in a stream can be used as an indication that the river is healthy. In Alaska, eight families including seventy-two species have been identified.

Willowflies Family Taeniopterygidae

***Taenionema* sp.**
Length: 3/8". Winter stoneflies are small, their bodies are dark and they emerge in late winter or in the spring. The nymphs live in large rivers. The adults feed on flowers. It is one of the most common stoneflies in south-central Alaska.

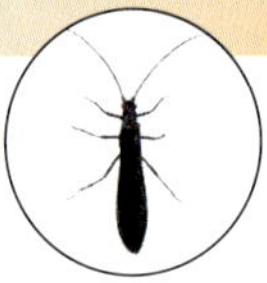

Perlodid Stoneflies Family Perlodidae

***Isoperla* sp.**
Length: 3/8" to 1". Body is blackish. Perlodid stoneflies emerge in the spring. The nymphs are usually found under large stones in medium to large rivers.

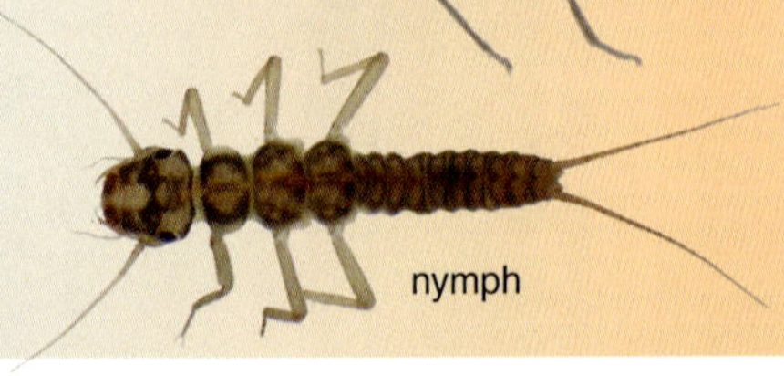

nymph

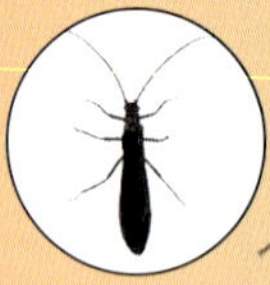

CADDISFLIES
ORDER TRICHOPTERA

Caddisfly adults look like drab moths measuring an inch or less. They are mostly nocturnal and can be found during the day resting on vegetation, their wings folded roof-like on their back.

Caddisfly pupa

The aquatic larvae live hidden inside shelters they fabricate out of twigs, coarse sand, spruce needles and other material available in their habitat. Some shelters are fastened to rocks, while others are mobile, hollow cylinders dragged around by the larvae as a hermit crab does with its borrowed snail shell. The moving cases can often be noticed on muddy lake bottoms or hanging from aquatic vegetation. Mature nymphs exit the pupal case and swim to the surface, crawl out from the water onto the vegetation, and emerge as adults. Caddisfly larvae are found in all freshwater habitats, with each species occupying a specific niche. This intimate relationship of the caddisfly larvae with their habitat makes them useful for monitoring the health of streams.

There are seventeen families and more than one hundred and eighty species in Alaska.

Family Molannidae

Molannodes tinctus
Length of the case: 5/8". Broad flanges extend on all sides around the chamber and completely cover the larva. The flattened cases are made out of fine rock. The larvae are found on muddy or sandy bottoms of lakes or slow rivers. They feed on organic particles and diatoms. There are six species in this genus in North America.

Family Rhyacophilidae

***Rhyacophila* sp.**
Larva length: up to 1" long
These larvae live in the interstices between boulders in fast running waters. Unlike the majority of caddisflies, *Rhyacophila* does not build a case. This genus comprises more than five hundred species worldwide found only in cool, fast running creeks.

Family Phryganeidae

The larva can be more than 1-1/2" long. In North America there are ten genera and twenty-eight species.

Phryganea cinerea

Adult length: 1". This large grey species is found mostly in marshes and lake margins. The larvae feed on both animal and vegetal matter and make their cases out of elongated shredded plant material organized in a spiral. This common species occurs throughout North America.

Family Brachycentridae

Found in fast flowing water, cases square in cross section are typical for some genera. Five genera and 30 species are found in North America

***Brachycentrus* sp.**

Larva length: 3/8". The larval cases are square in cross section and tapered. They are constructed of short pieces of grass organized transversely. The larvae live only in running creeks and they feed by filtering the water in the fast current. *Brachycentrus* attaches the ventral end of its case to a hard substrate with silk.

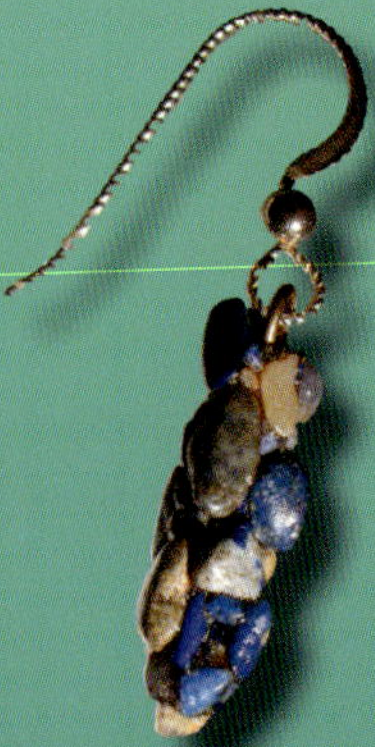

Caddisfly jewelry

A couple of enterprising biologists collect young caddisfly larvae and rear them in an artificial environment where the gravel river bottom has been replaced by crushed and polished semi-precious stones. The captive caddisfly larvae incorporate the substituted semi-precious stones to their cases, which will be collected after the adults emerge. The adorned case is then used as jewelry. See Wildscape.com.

Family Limnephilidae

Nemotaulius hostilis
Adult length: 1". The larvae live in well-vegetated ponds. The cases are fabricated with sections of leaves arranged laterally, giving the case a flattened appearance. This large drab brown species is widely distributed in North America.

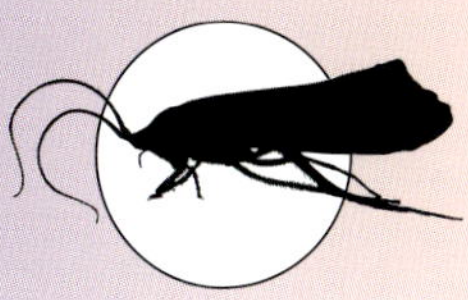

***Limnephilus* sp.**
The larvae live in various fresh water habitats. Larvae of this genus make their cases out of a variety of materials.

Examples of ***Limnephilidae***

These live in dense vegetation of pools and streams. The case is made of grass and sedge pieces organized transversely.

BEETLES
ORDER COLEOPTERA
Leaf Beetles
Family Chrysomelidae

Both adults and larvae of each species usually feed on one or a few closely related plant species. Generalist predators such as Coccinellidae, Vespidae, and spiders prey on leaf beetle larvae. When threatened, adults release their grip, curl up and fall to the ground. Aesthetic damage from larval feeding on willows and poplars can be considerable. Forty-two species in thirteen genera are known from Alaska.

Lady beetle for comparison

Freshly emerged adult

Chrysomela aeneicollis
Adult coloration varies greatly within the same colony.

Willow Leaf Beetles
***Chrysomela aeneicollis* species-group**
Length: 1/4". It is not clear at this point if the three species known from south-central Alaska, *Chrysomela falsa, C. aeneicollis* and *C. interna,* should be all synonymized into a single species as *C. aeneicollis.* These leaf beetles are ladybeetle-like, brightly colored red, orange or light brown with or without black markings or almost entirely black. The patterns and the dominant background vary considerably among individuals of the same colony. The long antennae distinguish them from lady beetles, which have short club-shaped antennae.

They are found in woodlands, parklands, and disturbed sites where host plants grow. Both the adults and larvae of willow leaf beetles feed on leaves of willows, cottonwood, balsam poplar and occasionally alder. They can be abundant on shrubs, forming loose colonies, but are not socially organized like ants or wasps.

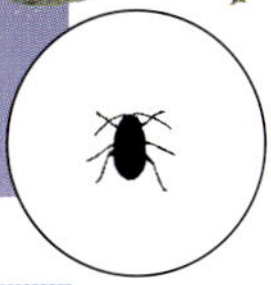

Phratora interstitialis

Length: 1/8" to 3/16". Metallic black-purplish or dark blue. Adults hide in the daytime with their head wedged in the crotch formed by the petiole (base of the leaves) and the stem. The larvae of *Phratora interstitialis* feed hidden under the dense hairs covering the underside of the leaves of feltleaf willow and Sitka willow. Three species occur in south-central Alaska. *P. kenaiensis* is found on aspen, *P. interstitialis* on Sitka willow, feltleaf willow, and *P. hudsonia* on paper birch. The genus *Phratora* is found throughout the northern hemisphere.

Bromius obscurus

Length: 3/16". Body dull brown or blackish, covered with dense short hairs, claws double pointed. Found in open woodlands and gardens. The beetle feeds on a variety of plants such as fireweed, strawberry and Scouler willow. Widespread throughout the northern hemisphere.

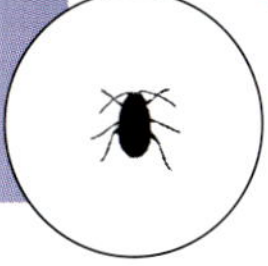

Flea Beetle
Altica tombacina

Length: 1/8". Dark metallic color, hind legs are strongly developed and adapted for jumping. Four species are known in Alaska. They are found in open woodlands and gardens and are known to feed on alder, birch, dogwood, rose, heuchera, strawberry, potentilla and fireweed. *Altica bimarginata* is metallic blue.

Phyllotreta zimmermanni

Length: 1/16". Elytra with black and yellow stripes. In Europe, *Phyllotreta undulata* causes considerable damage to vegetables, especially in moist sites. *Phyllotreta* is found throughout the northern hemisphere.

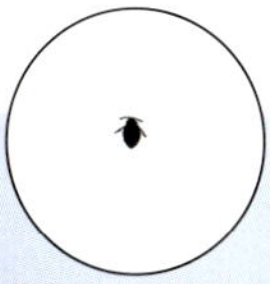

Hippuriphila mancula

Length: 1/16". Elytra bronze with tips yellow. It is found in moist places. Adults feed on horsetail. A second species, *H. canadensis,* is also found in Alaska. *H. mancula* is found throughout the northern hemisphere.

Aquatic and Semi-aquatic Leaf Beetles

The body of aquatic species is covered with dense water-repellent hairs that trap a layer of air for breathing during submersion. Larvae of aquatic species feed mostly on submerged vegetation while adults feed above the surface.

***Plateumaris* sp.**

Length: 3/8". Body is metallic colored and coarsely sculptured, and the antennae are long. In the spring, the females lay eggs on submerged portions of the food plant. The aquatic larvae feed on plant sap and get their oxygen by boring into hollow plant cells that store oxygen, using two hooks at the end of the abdomen. At the end of the summer, the larvae pupate inside submerged cocoons filled with air. The adults metamorphose in the fall but remain in the cocoon until spring. Beetles of this genus, of which 6 species are known from Alaska, live on a variety of plants growing in wetlands and at the margin of lakes, such as rush, bulrush, sedge, calamus, spikerush, marsh marigold and buttercup. *Plateumaris* is found throughout the northern hemisphere.

Galerucella nymphaeae

Length: 3/8". Both larvae and adults are semi-aquatic and feed on floating vegetation in lakes and ponds such as yellow pond lily, polygonum, sweet gale, dock and pondweed. Beetle browsing causes brown blotches on the leaves. One species is known to be in Alaska. The genus is represented throughout the northern hemisphere.

Lady Beetles
Family Coccinellidae

Orange lady beetles are very popular. Their "cuteness" some say, comes from the two white spots on either side of the thorax that mimic big infant eyes. The orange hemispherical shape is also attractive. Several species are commercially raised for controlling aphids in greenhouses and gardens. This family is represented in Alaska by twelve genera with twenty-six species.

Transverse Lady Beetle
Coccinella transversoguttata

Length: 1/4". The name refers to the transverse black spots on the elytra. This distinctive beetle is a voracious predator of aphids, both as adults and as a larvae.

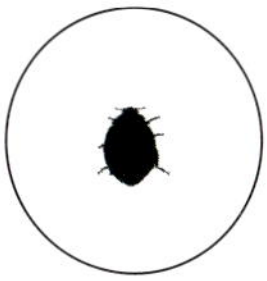

Three-banded Lady Beetle
Coccinella trifasciata

Length: 3/16". During the year of its birth, this beetle is yellow with black transversal banding. The following spring, after hibernation, the yellow turns orange-red except for a pale halo around the black markings.

First year

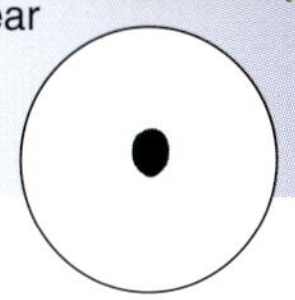

Second year

Eyed-spotted Lady Beetle
Anatis mali

Length: 1/4". The black spots with yellow rings are the best field marks. This species is a generalist predator of soft-bodied insects, from spruce budworms to aphids. A study showed that about 270 aphids were consumed by a single individual during its larval development.

Twelve-spotted Lady Beetle
Calvia quatuordecimguttata
Length: 3/16". Broadly distributed in North America.

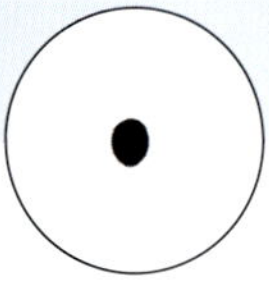

Two similar looking beetle larvae, two different life histories:

Lady Beetle larva

Lady beetle adults and larvae are voracious predators of aphids. Adults and larvae tend to be solitary.

Willow Leaf Beetle larva

Willow leaf beetle adults and larvae eat willow and cotton-wood leaves. Adults and larvae tend to live in groups.

Ground Beetles
Family Carabidae

The antennae are long and wiry, the mandibles are well developed, and the legs long and slender, and well articulated, making them adapted for fast running or digging. The elytra are usually hard. Ground beetles live mostly in the leaf litter or in open areas and most, with the exception of *Cicindella* and *Bembidion,* are poor fliers. Some species of *Carabus, Pterostichus*, and *Nebria* have reduced flying wings or flying muscles and are subsequently flightless. Most ground beetles are generalist predators, though a few are omnivorous. Most larvae are carnivorous. The family is represented in Alaska by more than two hundred fifty species.

Snail Eater
Scaphinotus marginatus
Length: 1/2" to 3/4". Mouth parts protruding forward, likely an adaptation to feed on snails. Found in the rich humus of coastal forests and under alder thickets. The beetle is nocturnal in its habits. *S. marginatus* is also found above tree line, near snow patches. Another species, *S. angusticollis,* is larger- up to 1"- and its elytra are almost smooth, velvet-like. *S. angusticollis* is found only in coastal areas. The genus is represented by many species throughout North America.

Carabus taedatus
Length: 5/8" to 1". Large species with long legs and stout mandibles. The elytra are sculpted with ridges and pits, not simple rows of punctures. Two other species are also found in south-central Alaska: *C. chamissonis* and *C. maeander.*

C. chamissonis is found in open dry country such as tundra, *C. maeander* is common in open moist vegetated habitats, such as sedge bogs, and *C. taedatus* can be found in open dry gravely soil. Species of this genus are widespread throughout the northern hemisphere.

Pterostichus sp.

Length: 3/16" to 5/8". This predatory beetle is stout with fairly short legs. *Pterostichus* live in various habitats ranging from tundra to forest. Ten species have been identified in south-central Alaska. *Pterostichus* species are found throughout the northern hemisphere.

Calathus ingratus

Length: 3/8". The front of the body is almost flat becoming slightly higher toward the rear, shaped like a wedge, which allows this beetle to force its way between layers of dead leaves in order to reach well-hidden prey. The entire body is reddish-brown with a translucent thorax margin. This beetle is found under dead leaves beneath deciduous shrubs, and in grasslands. *C. advena* is the only species of the genus found in Alaska. The genus is found throughout the northern hemisphere.

Elaphrus purpurans

Length: 1/4". This beetle is similar to the tiger beetle, but is much smaller and is not as brightly colored. The eyes are bulging and the elytra are a dull greenish color with rows of three shiny raised "mirrors." *Elaphrus* live on sandy riverbanks of fast-moving rivers where it runs about in the daytime in search of terrestrial insects that have become ensnared within the water surface. Carabids of the genus *Opisthius,* resembling *Elaphrus* and found in similar habitat, are larger (3/8") and more oval. Nine species of *Elaphrus* have been recorded in Alaska. The genus is represented throughout the northern hemisphere.

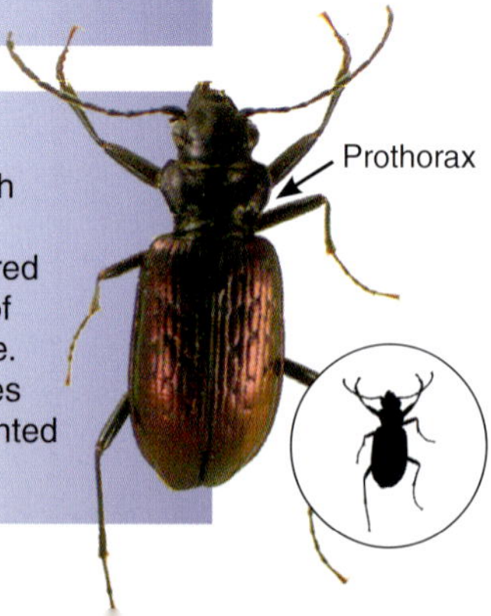

Nebria metallica

Length: 1/4" to 5/8". *Nebria* has a short prothorax with strongly double S-shaped sides, and the elytra are sometimes metallic, and of a single color. It has two red spots on the forehead. It is found in barren margins of running water, and often in snowfields above tree line. *Nebria* is nocturnal and carnivorous. Fourteen species have been identified in Alaska. The genus is represented throughout the northern hemisphere.

Amara littoralis
Length 1/4". A bronze-colored compact beetle, very common in disturbed habitats found around houses or near trails. It often climbs on vegetation and is an herbivore that feeds on seeds. Thirteen species of *Amara* are found in south-central Alaska.

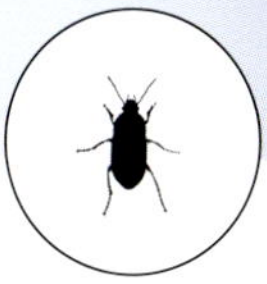

Amara hyperborea
Length: 3/8". This species lives in dry sandy soils, often in dead leaves of alders or willows. *A. hyperborea* is found in the northern conifer forest, along lakes and ocean shores. The species is found throughout the northern hemisphere.

Bembidion spp.
Length: 1/8" to 1/4". These are small, swift beetles, with long legs and antennae. Their elytra are often metallic and spotted; the terminal palpus segment is reduced. Most species are found close to water, feeding on dying insects drifting ashore. Other species are found in the lush vegetation of marshes, or in dry soils devoid of vegetation. Twenty-nine species have been identified in Alaska. The genus is well represented throughout the northern hemisphere.

Notiophilus semistriatus
Length: less than 1/4". The genus can be recognized by the bulging eyes. They are predators and three species are reported to be in south-central Alaska. *N. semistriatus* is found throughout North America.

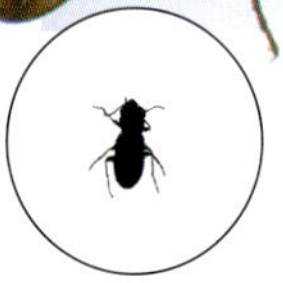

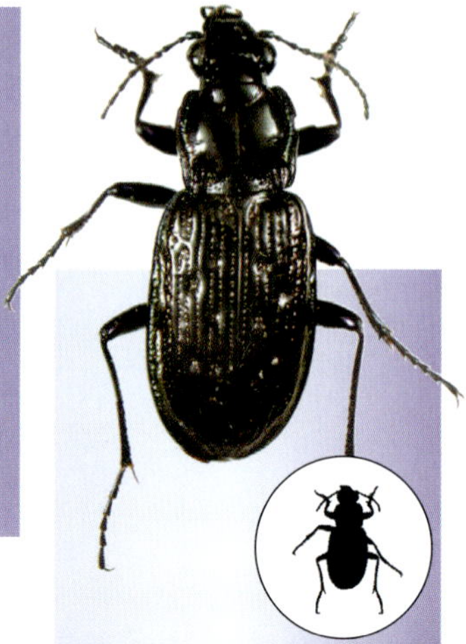

Blethisa quadricollis
Length: 1/2". The larva is partly aquatic. This species lives in sphagnum moss mounts, and in acid bogs at the margin of ponds. It is often associated with bog cranberry and sweet gale. This insect hibernates as an adult. One species is known in south-central Alaska, and four are known to be in Alaska. *B. quadricollis* is found throughout North America.

Dyschirius truncatus
Length: less than 1/8". This is one of two species known to be in south-central Alaska. This minute carabid is found in very wet habitats, often on barren clays along shorelines or riverbanks. It is found throughout most of North America.

Tiger Beetle
Cicindela oregona
Length: 1/2". Ferocious-looking beetle, armed with long sickle-shaped mandibles. The elytra are bright green with a black and yellow pattern. The underside of the body is covered with dense white hairs. *Cicindela* is found on barren sandbars on seashores and on the low banks of wide rivers. The beetles can be numerous in the proper habitat. The adult beetles hunt actively during warm sunny days, pouncing on their prey. When disturbed, the tiger beetles swiftly take off and land a few dozen feet away. The larvae stalk their prey from the opening of a vertical shaft dug in the soil. Three species are known to be in Alaska and the genus is found throughout the northern hemisphere.

Carrion Beetles
Family Silphidae

In south-central Alaska, eight species are known from three genera in this family.

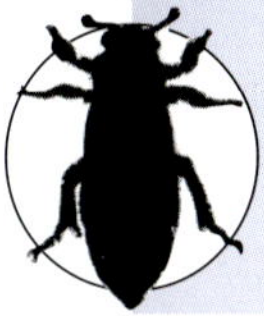

Carrion Beetle
Nicrophorus investigator

Length: 1/2" to 3/4". This brightly colored beetle, large by Alaskan insect standards, uses carcasses of small birds or mammals to feed its offspring. It is attracted by the odor of carrion, which it can locate from quite a distance. Several carrion beetles may reach a carcass but usually only one pair claims the remains, chasing off the competitors. After the carcass is buried, the female digs a chamber in the soil nearby in which it lays some 30 eggs. The female remains near the carcass and feeds the young larvae. When disturbed, the beetle produces an audible stridulation, that sounds like a bumble bee. The contrasting red and black pattern warns predators against the foul smelling secretion the beetle discharges when disturbed. *Nicrophorus* are found throughout the northern hemisphere.

Carrion beetles are easily baited with a piece of meat left in an open can partly buried in the soil.

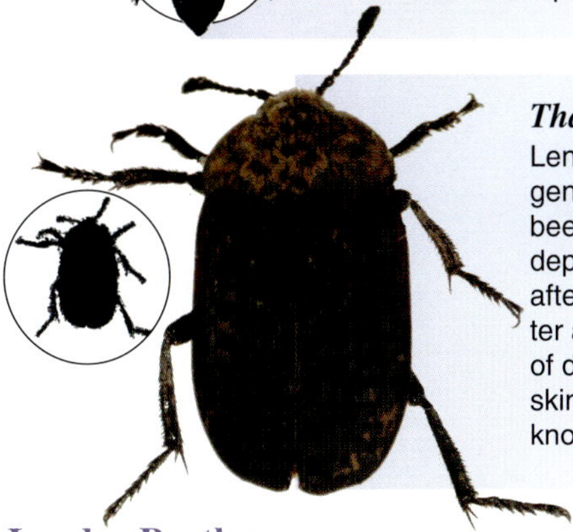

Thanatophilus lapponicus

Length: 3/8". One of two species of this genus found in south-central Alaska. These beetles search for carcasses and the female deposits 10 or so eggs near it. Adults mature after more than a month and they overwinter as adults. These beetles can be a pest of dried animal products such as untanned skins, dried meat or fish. This species is known throughout the northern hemisphere.

Larder Beetles
Family Dermestidae

Dermestes lardarius

Length: 3/8". Larder beetles eat old dried meat. They are found infesting stored dog food, untanned furs and food pantries where they feed on spilled food. The larder beetles are used by taxidermists for cleaning skulls and bones for museum collections and they are the bane of insect collections where they do damage to pinned insect specimens. Two species can be found in Alaska.

Predaceous Diving Beetles
Family Dytiscidae

Ninety-four species of Dytiscidae known to be in Alaska are distributed in thirteen genera.

***Dytiscus* sp.**

Length: up to 1-3/4". These large diving beetles prey on small fish, tadpoles and insects in lakes and ponds. Males and females can easily be distinguished by the longitudinal grooves of the elytra of the female and the smooth and polished male elytra. The male's forelegs bear distinctive suction cups. The larvae look ferocious with their large hooked piercing mandibles with which they paralyze their prey by injection of a venom. This venom liquefies the tissue that is then sucked back by the larvae. All that remains of the prey is an empty skin or carapace. *Dytiscus* breathe oxygen from the air which is stored in the space between the elytra and abdomen.

In midsummer, the adults crawl out of the water and fly, dispersing in search of new habitat.

Dytiscus dauricus

Whirligig Beetles
Family Gyrinidae

This family is represented in Alaska by the genus *Gyrinus*.

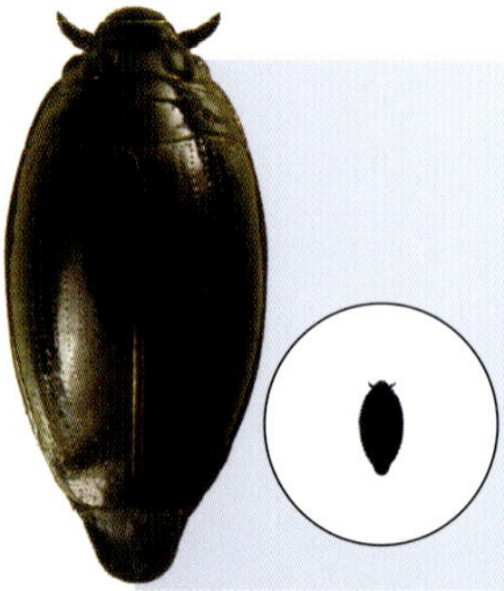

***Gyrinus* sp.**

Length: 1/4". The adults of this genus rapidly whirl in loose groups on the surface of lakes, pools or in river eddies often near the shore. These beetles have eyes curiously divided into two with the lower part adapted to see underwater and the upper half able to see above the insect. The larvae are aquatic and pupation occurs outside the water in river bank vegetation. Seven species of this genus are known to be in Alaska.

Longhorned Beetles
Family Cerambycidae

The antennae are at least half the length of the body. Cerambycid larvae are wood borers and adults of some species can be found on flowers. The family has more than 31 species in Alaska.

***Saperda* sp.**
Length: 3/8". Adults are black and covered with short grey hairs. The females gnaw a notch in the bark and deposit 1 or 2 eggs per notch. The young larvae bore first under the bark, then into the wood itself. Adults emerge in May. In south-central Alaska, this genus has been found on older feltleaf willow in the Matanuska valley.

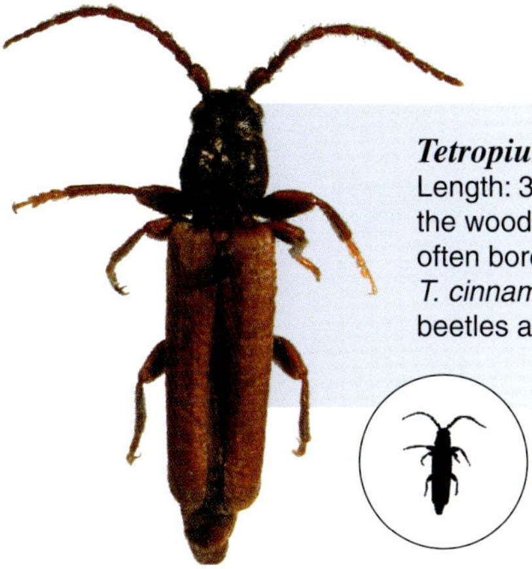

Tetropium cinnamopterum
Length: 3/8". The larva bores between the bark and the wood of dead or freshly killed spruce trees. It often bores a chamber in the wood to pupate. *T. cinnamopterum* is often found alongside bark beetles and can be found throughout North America.

Pronocera collaris
Length: 3/8". The larva of this species lives in the same phloem layer as the bark beetles.

Xestoleptura behrensi
Length: 3/8" to 1/2". Adults can be found feeding on the flowers of cow parsnip and other meadow flowers. The genus is found throughout the northern hemisphere.

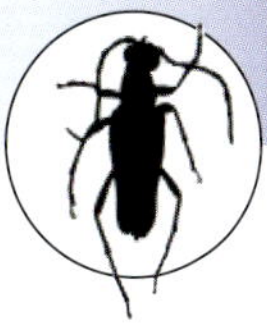

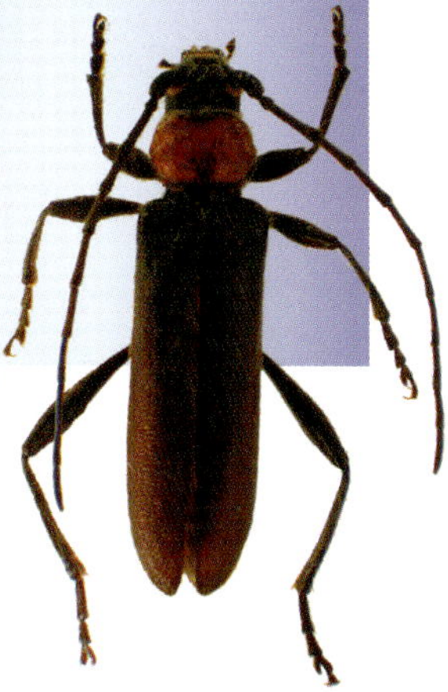

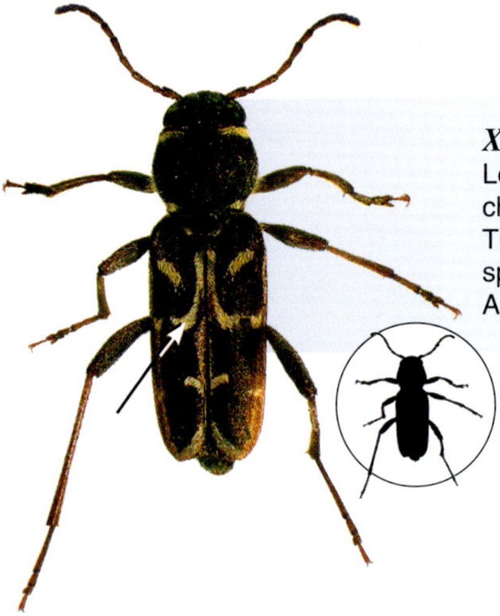

Xylotrechus undulatus
Length: 3/8" to 1/2". Body hairy, yellow chevron pattern on a dark background. The larva bores into the main stem of dead spruce and larch. Two species are known for Alaska; common in northwestern America.

Pogonocherus penicillatus
Length: 3/8". The larva bores into conifers and one species are known to occur in Alaska.

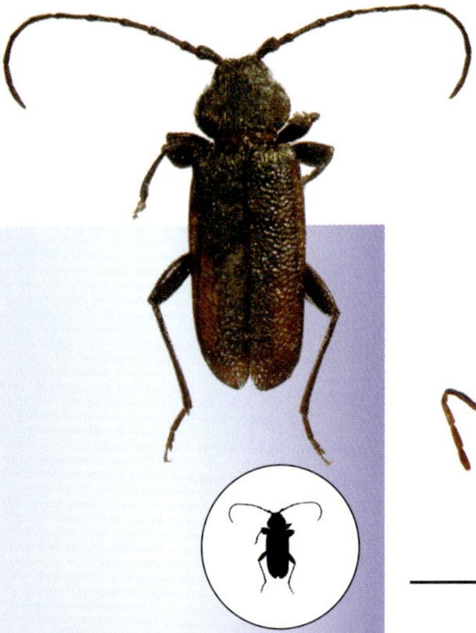

New House Borer
Arhopalus cicatrosum
Length: 1/4". The larvae of this species bore into freshly fire-killed spruce. Occasionally, freshly cut logs used to build homes contain larvae that will emerge in the new building, hence the name. One species is known in Alaska.

Phymatodes dimidiatus
Length: 1/4". The larva lives in small spruce or larch branches or logs with thin bark. It is found throughout western North America, one species is known to be in Alaska.

White-spotted Sawyer
Monochamus scutellatus
Body length: 3/4". This large species is common in south-central Alaska. The larvae bore into the trunks of dying or dead spruce. Adults feed on the bark on the underside of twigs. The female lays eggs in crevices that they chew in the bark. The larva, at first, mines under the bark, then deeper into the wood itself, then back near the surface of the trunk prior to pupating. The larva takes one or two years to mature the emergence hole is perfectly round.

Adults are attracted by the smoke of forest fires. This species is found throughout North America

Grammoptera subargentata
Length: 1/8". The adult body is black, densely hairy above and below. The larvae live in poplar wood. The adults feed on flowers.

Judolia montivagans
Length: 3/8". The larvae bore into spruce, willow and aspen. Adults feed on flowers. This species is found throughout North America.

Pachyta lamed
Length: 1/2". The larva is found in the trunk of spruce trees. Adults feed on cow parsnip flowers. The larva takes two years to mature. This species is found in the northern region of the northern hemisphere.

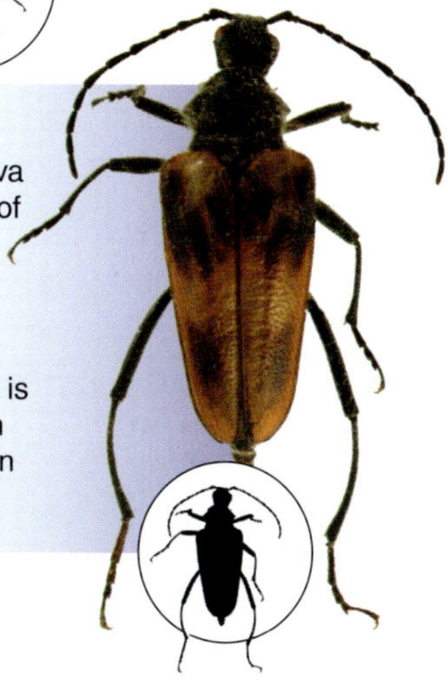

Curculionidae

Recently, the family Scolytidae has been changed to Scolytinae, a subfamily of Curculionidae.

Weevil larva

Weevils, Snout Beetles
Subfamily Curculioninae

Weevils have a characteristic snout, their antennae are elbowed and their body is more or less covered with fine scales. The elytra are very hard and the body is often pear-shaped. Weevils are herbivores usually specializing in a few food plants. The larvae live inside the plants or underground, feeding on roots.

Some twenty genera and sixty-one species have been identified in Alaska.

Willow Weevil
Dorytomus leucophyllus
Length: 1/16". All the weevils of this genus live on willows, poplars and cottonwoods. Most of the larvae live in catkins and one species lives in sawfly and gall midge galls. The larvae develop in the catkins, then drop to the ground where they pupate. The adults emerge in mid-summer, and mate late in the summer. The females oviposit in the following year's catkin buds.

Grypus equiseti
Length: 1/4". The larva of this long-snouted weevil bores into the base of the stem of horsetail stem. The adult overwinters. This species is found throughout the northern hemisphere.

Trichalophus alternatus
Length: 3/8". This hard-shelled, pear-shaped weevil is very common. The elytra have a bronze satin finish. The larva likely feeds on the roots of shrubs. The adult, which is unable to fly, is found on the foliage of Scouler, grey-leafed and bebb willows as well as ornamental ash. This snout beetle occasionally enters homes at the end of the summer.

Sthereus ptinoides
Length: 1/4". This species lives in grassy coastal meadows and hides under driftwood in areas flooded regularly at high tide. The coarse-textured elytra are encrusted with silt deposited while the insect is submerged.

Cylindrical Root Weevil
Rhyncolus brunneus
Length: 1/8". This common weevil is found in older forests and is found in old decaying roots and in decaying wood structures. This species is found throughout western North America.

Lateral view

Lepyrus gemellus
Length: 1/2". Pear-shaped beetle, hard shelled, black with longitudinal white stripes, not shiny. The adults can be found throughout the summer on Sitka, Barclay, feltleaf and Hooker willows but the larvae probably live on other plants.

Bark and Ambrogia Beetles
Subfamily Scolytinae

Some twenty-six species are known from south-central Alaska, most of which live under the bark of spruce trees. Most of these beetles live in the phloem or inner cambium of the bark. *Ambrosia* beetles bore into the sapwood.

Spruce Beetle
Dendroctonus rufipennis

Length: 1/4". This species is the largest of the bark beetles in south-central Alaska. The species bores into the phloem layer of the main trunk of white, Sitka, and Lutz spruce. This species is responsible for the deforestation of three million acres in south-central Alaska since the 1990s.

Four-eyed Spruce Beetle
Polygraphus rufipennis

Length: 1/16". The beetle has a dark body color with yellowish hairs. The eyes are split in two. This species is adapted to live under the bark of trees that have burned or are attacked by other bark beetles. It is known from coniferous forests throughout borcal regions of North America.

EngraverBeetles
Ips tridens and
Ips pertusbatus

Length: 1/8". The rear of beetles in this genus is concave and the lateral edge of the recess is ornate with multiple spines. *Ips* beetles attack mostly fallen trees. There are four species of *Ips* recognized in south-central Alaska. The genus is found throughout the northern hemisphere.

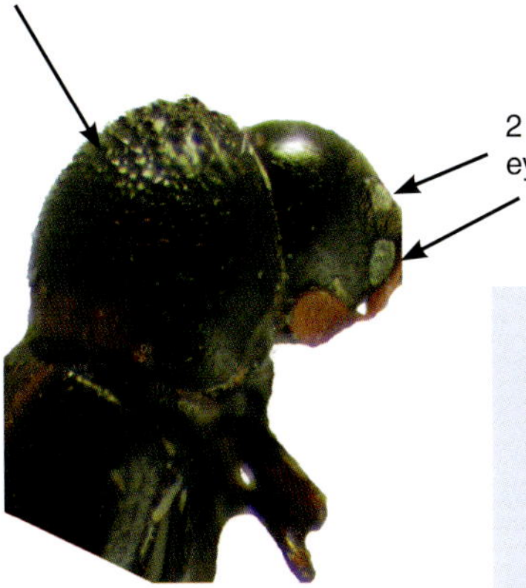

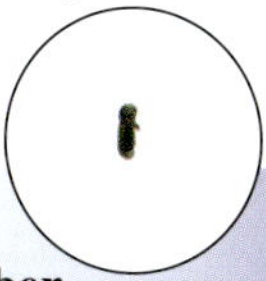

***Dryocoetes* affaber**

Length: 1/8". Females have a dense brush of yellow hair on the front of the head. This beetle attacks fallen trees and lower sections of dying spruce and are numerous after a forest fire. Three species are known to be in south-central Alaska and ninety species have been described worldwide.

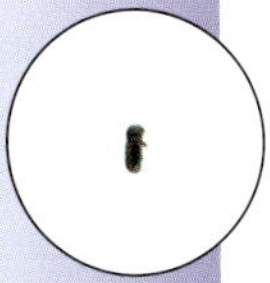

Striped Ambrosia Beetle
Trypodendron lineatum

Length: 1/8". The head of this insect faces downward, suggesting a ball and socket joint. These beetles also have two distinct pairs of eyes and the back is dark brown with longitudinal light-colored stripes. Four species are known to occur in Alaska. The larvae of two live on spruce; the two other species live on birch, alder, and poplar. This beetle inoculates the wood of dying spruce with a fungus on which it feeds. There are about twelve species in this genus throughout the northern hemisphere.

Sylvanid Beetles
Family Sylvanidae

Dendrophagus cygnaei

Length: 1/4". This beetle has a very flat profile and is a likely predator of bark beetles.

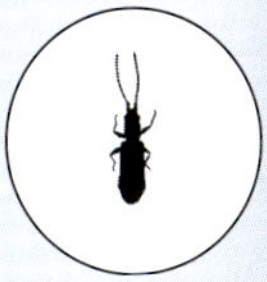

Scarab Beetles
Family Scarabaeidae

Aphodius leopardus
Length: 1/4". The short comb-like antennae and the broad spiny fore-tibia are distinctive for the family. *Aphodius* is a very large genus (more than one thousand species worldwide with eight in Alaska). *Aphodius* are found in the droppings of large herbivorous mammals. Each species of *Aphodius* seems to prefer specific mammals. The female digs a tunnel under the dung, shapes some dung into a ball, brings it down into the tunnel and lays eggs on it.

Aegialia montana

Length: 3/16". This genus is represented in Alaska by four species.

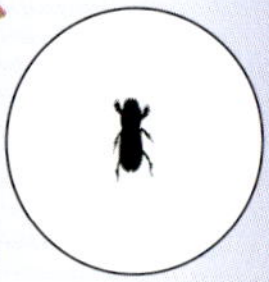

Variegated Mud-loving Beetles
Family Heteroceridae

In Alaska, two genera each with a single species have been identified.

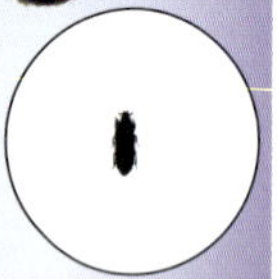

Length 1/8". These small beetles are found in water saturated soil, moss and on the banks of creeks.

Checkered Beetles
Family Cleridae

In South Central Alaska, the Cleridae family includes four species in three genera. The last three antennal segments are club-shaped.

Thanasimus nubilus
Length: 1/4". The larvae live under spruce bark where they hunt bark beetle larvae. This species is often captured in pheromone-lure traps used to monitor bark beetle populations. Three species *T. nubilus* and *T. undulatus* occur naturally in Alaska species.

Pythid Beetles
Family Pythidae

The larvae live under the bark of conifer trees. Three species are known from Alaska.

Priognathus monilicornis
Length: 3/8". The larvae are likely bark beetle predators.

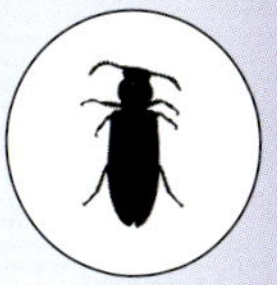

Ant-like Flower Beetles
Family Anthicidae

Anthicus coracinus
Length: 3/16". The thorax is narrow at the rear. These beetles are omnivorous, opportunistic predators. Adults are protected by a chemical and can venture into ant nests without being attacked.

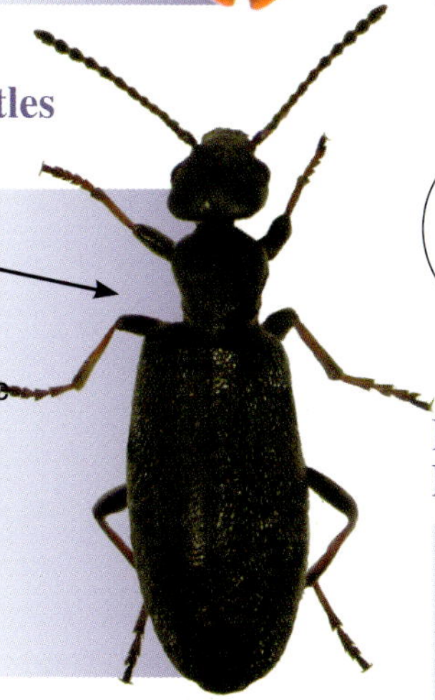

Fireflies
Family Lampyridae

***Ellychnia* sp.**
Length: 5/16". The head is not visible from above and there is a “window”, (unpigmented area) in the thorax extension above each eye. The larvae feed on smaller insects and slugs. Adults of these species of fireflies do not have light-producing organs.

Narrow-waisted Bark Beetles
Family Salpingidae

Three species are known to be within Alaska.

Aegialites alaskensis
Length: 1/8" to 3/16". These insects live along the ocean shoreline, and can be found within the beach rye belt or among rocks.

Round Fungus Beetles
Family Leiodidae

Length: less than 3/32". Adults and larvae feed on fungi. For defense, the adult will roll into a ball. There are about forty species known from Alaska.

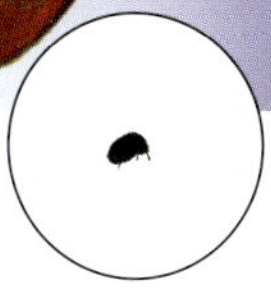

Net-winged Beetles
Family Lycidae

Dictyopterus aurora
Length: 3/8" to 1/2". Bright red elytra with conspicuous ridges and sculptures on the thorax. Both adults and larvae are predators. Eggs are deposited in decaying tree stumps. The bright red color warns predators that the species is toxic.

This species is one of the few northern representative of the family Lycidae that is so abundant in the tropics.

Fire-colored Beetles
Family Pyrochroidae

Dendroides ephemeroides
Length: 1/2". Flat elongated orange beetle, males have distinctive branched antennae. The larvae live in decaying wood and feed on fungi and molds. Adults do not eat.

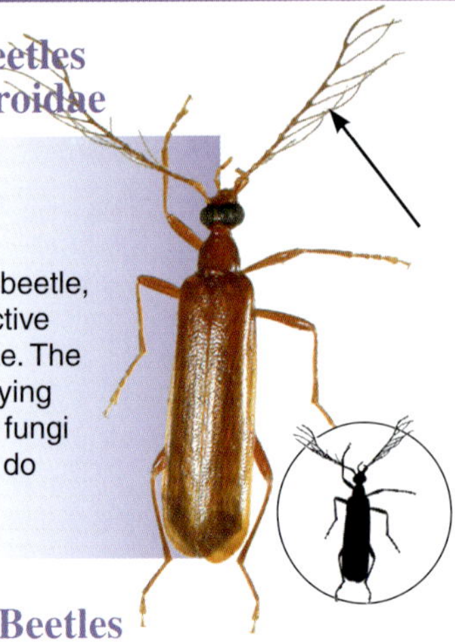

Bark-gnawing Beetles
Family Trogositidae

Ostoma columbiana
Length: 3/8". This species feeds on fungi in decaying wood. Another species, *O. ferruginea* is slightly larger and entirely dark brown with a slightly lighter rim. It is found in decaying birch.

Flat Bark Beetles
Family Cucujidae

Cucujus clavipes
Length: 1/2". Bright red flattened beetles. The adults and larvae live in the galleries of wood borers and bark beetles, which they hunt.

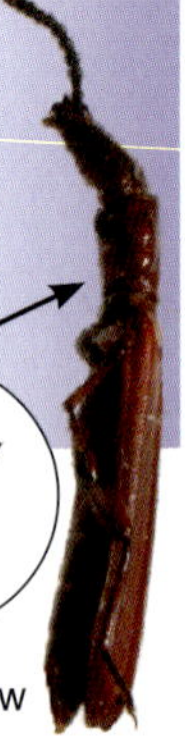

Sawtoothed Grain Beetle
Oryzaephilus surinamensis
Length: 1/8". Common in pantries and often found in flour and grain products. This minute insect can be a serious pest as both the larvae and the adults can be found in packages of dried products left open for long periods of time. They are originally from tropical regions.

Metallic Wood boring Beetles
Family Buprestidae

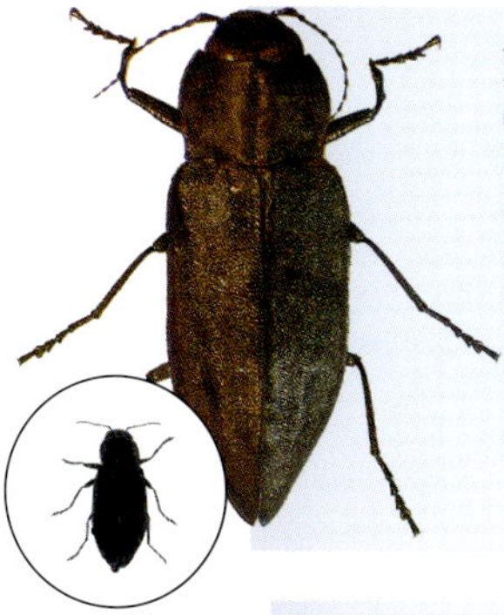

Flat-headed Borer
Oxypteris acuminata
Length: 3/8". In south-central Alaska the larva of this species bores into the bark of dying or dead spruce. It is found in most of the northern hemisphere.

***Dicerca* sp.**
Length: 3/4". The genus can be recognized by the double pointed end of the elytra. The larva bores into the sapwood and heartwood of spruce and aspen. Three species are found in Alaska.

Melanophila fulvoguttata
Length 3/8". Females of this species deposit eggs in bark crevices of dying spruces. The larvae make winding galleries in the cambium. The life cycle lasts one or two years.

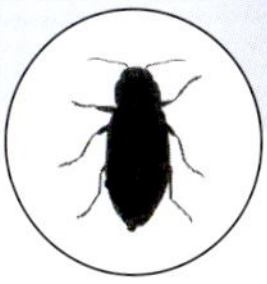

Click Beetles
Family Elateridae

The articulation between the thorax and the abdomen is equipped with a spring-loaded mechanism. When the beetle is held upside down on its back, it bends its thorax toward the belly, setting the articulation under tension. The sudden release of the mechanism projects the beetle several inches up in the air with an audible clicking sound. If the beetle does not land right side up, it repeats the maneuver.

***Ctenicera* sp.**

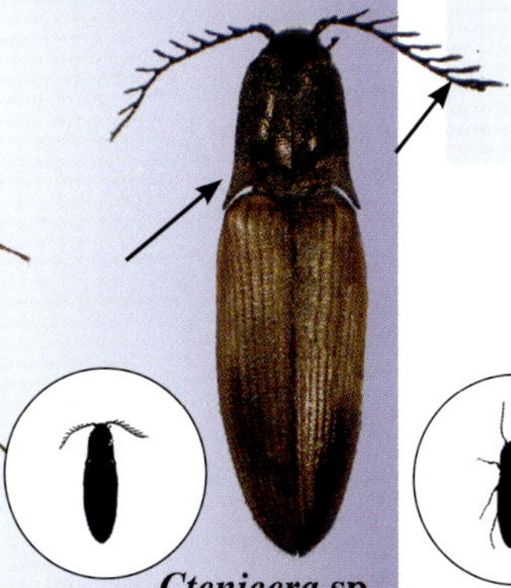

***Ctenicera* sp.**

Length: 3/8". The larva bores into the stems and the bark of dead or dying white spruce. It is found in most of North America. There are three species of this genus in Alaska.

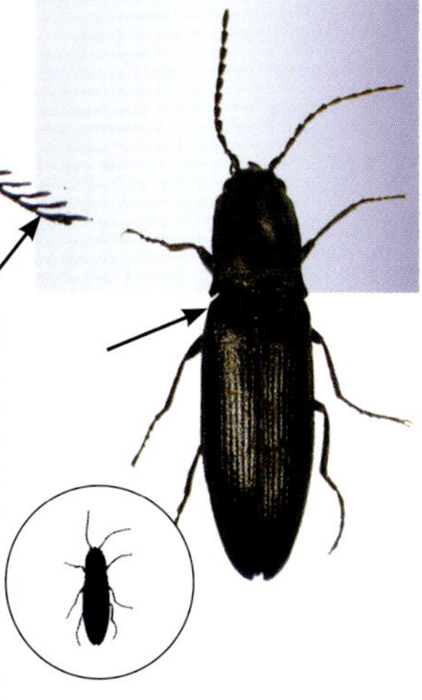

Darkling Beetles
Family Tenebrionidae

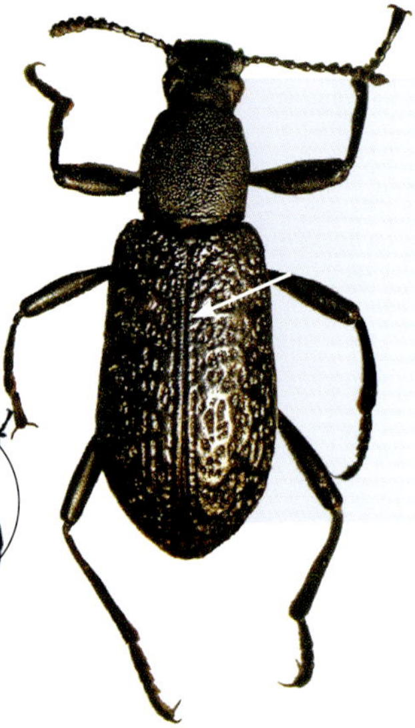

Upis ceramboides
Length: 5/8" to 3/4". This fair-sized beetle is easily identified by the coarse granulation of the elytra. The beetle can be found in galleries inside decaying birch trees. *Upis* have almost disappeared from southern Sweden because of the lack of decaying birches in the forests.

Corticiceus pratermissus
Length: 3/16". This species is associated with bark beetle galleries in conifers.

Cylindrical Bark Beetles
Family Colydiidae

Lasconotus borealis
Length: 1/8". The elytra of this beetle are longitudinally ridged and are velvety in appearance. *Lasconotus* is a bark beetle predator.

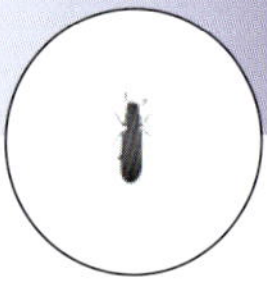

Spider Beetles
Family Anobiidae

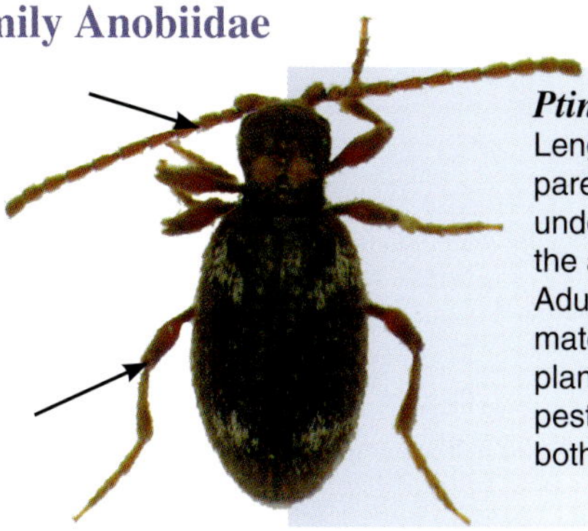

***Ptinus* sp.**
Length: 3/16". The thorax is narrow compared to the elytra and the head is hidden under the thorax when viewed from above; the antennae are long. The elytra are hairy. Adults and larvae feed on dry organic material such as animal carcasses, dried plants or dried food stuffs. They can be pests in stored dog food, and can be found both outside and inside buildings.

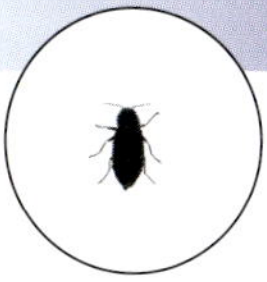

Minute Brown Scavenger Beetles
Family Latridiidae

Length: 1/16". The body of these beetles is widest at the middle, and the thorax is almost circular as seen from above. Antennae are clubbed and the elytra are marked with longitudinal rows of punctures. These beetles are associated with bark beetles, but the nature of the relationship is unknown. They are often found in decaying organic material and are believed to feed on fungi.

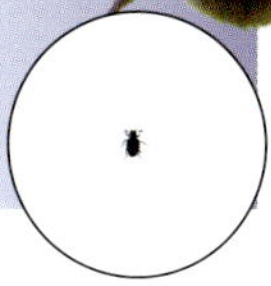

Silken Fungus Beetles
Family Cryptophagidae

Length: 1/8". Most of these beetles are found in old birds' nests and near mold or fungi. Sometimes they are found with stored food where they feed on mold. In Alaska, twenty-eight species are recognized within eight genera.

Rhizophagid Beetles
Family Monotomidae

Three species of *Rhizophagus* are found in Alaska, mostly in decaying vegetation.

Rhizophagus dimidiatus
Length: 1/8". These beetles live under the bark of dead trees where they hunt for wood boring insects. They are generalist predators.

Rove Beetles
Family Staphylinidae

Rove beetles are predators of maggots, worms and mites. They are found where their prey are numerous in decaying organic material such as carrion or compost. The entire family is quite distinctive, distinguished by short, square elytra that do not conceal their abdomen. A few other beetle families also have species with short elytra, but usually the abdomen is not as flexible as that of rove beetles. Earwigs (Forficulidae) have pincers at the end of the abdomen. More than 300 species of rove beetles have been identified in Alaska.

Large Hairy Rove Beetle
Creophilus maxillosus

Length: 5/8" to 3/4". The elytra of this beetle are very short and cover only the anterior section of the abdomen. The body is very hairy. The larvae of this large rove beetle are predators of carrion beetle larvae often found on animal carcasses.

***Tachyporus* sp.**

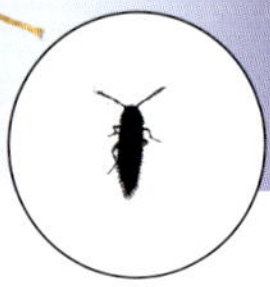

Shining Fungus Beetles
Family Scaphidiidae

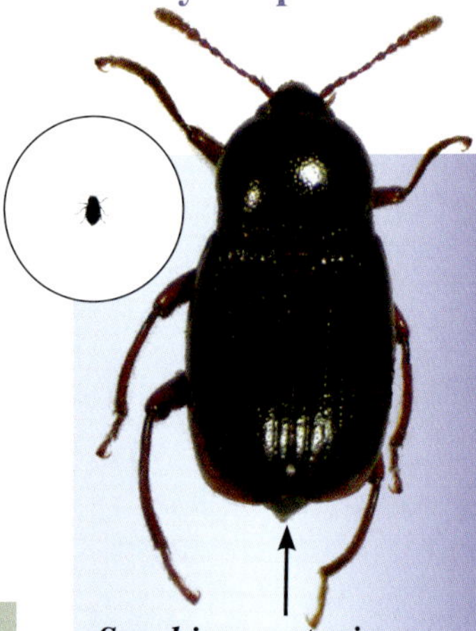

Scaphium castanipes

Length 3/16". A single species is known to be in Alaska. The small size, distinctive shape, and the pointed abdomen extending past the truncated elytra are characteristic for this species. These beetles are common under bark, and in the rich soil of the forest floor where they feed on mold and fungi.

Feather-winged Beetles
Family Ptiliidae

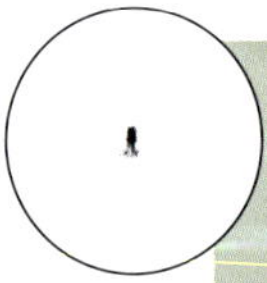

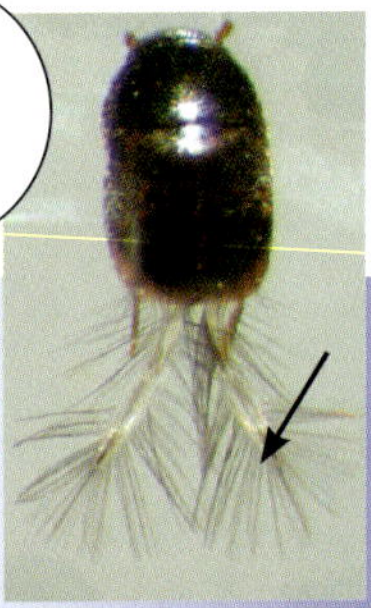

***Acrotrichis* sp.**

Length: less than 1/16". The wings of these very small insects are bordered with long straight hairs. Four species of the genus *Acrotrichis* are found in Alaska.

TWISTED-WINGED PARASITES
ORDER STREPSIPTERA

Females of this family are obligatory internal parasites of other insects. They remain partially inside their hosts when mating and releasing their larvae. For many species, females have no eyes, antennae, legs or external genitalia. Males are winged and they locate a female by the pheromone she releases. The closest relatives of Strepsiptera seem to be either Diptera or Coleoptera.

Family Elenchidae

Elenchus sp.
Length: less than 1/16". The males of these minute insects have only one pair of fan-shaped hindwings, and the forewings are reduced to stubs. They live less than one day as adults. The females are wingless and live partially as internal parasites in leaf hoppers, bees or wasps, with their cephalothorax protruding from the body of their host. The female releases pheromones that attract flying males. The female produces several thousand larvae while remaining inside the host. The minute larvae search on their own for a host to parasitize.

FLEAS
ORDER SIPHONAPTERA

Common Fleas
Family Pulicidae

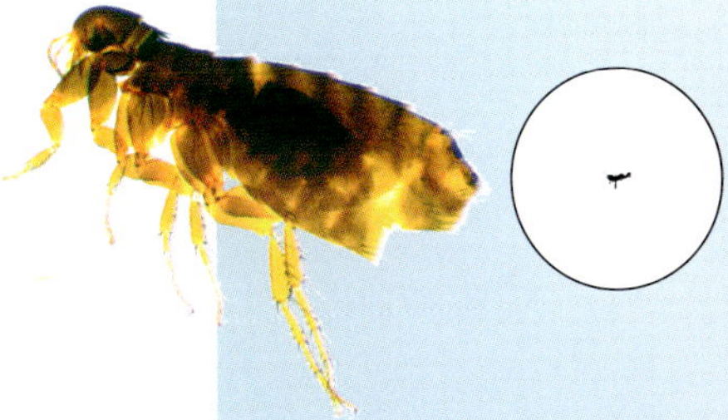

Length: less than 1/16". Fleas from this family parasitize humans and other mammals. They may change host species, from dog, rat or cats to humans. They are responsible for transmitting several diseases such as the bubonic plague, which was transmitted from rats to humans.

FLIES
ORDER DIPTERA

The term “fly” is used here in the broadest sense. It include the flies, mosquitoes, midges, etc. The most distinguishing characteristic for these insects is a pair of well-developed forewings and hindwings (halteres) reduced to stubs.

Mosquitoes
Family Culicidae

Males feed only on plant nectar. Females also feed on nectar but usually need a blood meal to mature their eggs. Larvae live in ponds and are most abundant in snowmelt temporary pools where there are no predatory fish. They take four days to a month to mature, depending on water temperature. Females have simple antennae and males have feathery antennae. Both sexes have a long proboscis. Twenty-seven species from four genera occur in Alaska. Four species of mosquitoes found in Alaska are known to be carriers of the West Nile Virus in the lower 48. The virus is transmitted from mosquitoes to birds and occasionally to mammals. The virus is deadly for birds but to date has not been found in Alaska.

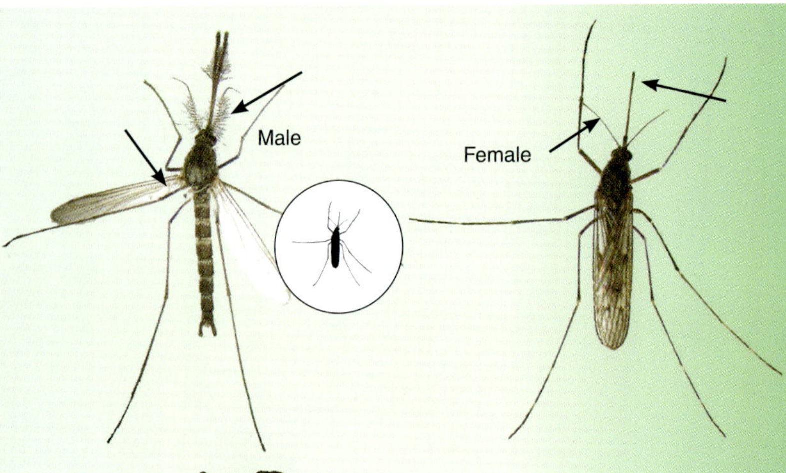

Culiseta incidens
Length: 1/4". Larvae live in ditches and semi-permanent streams and have one or more generations per year. The females overwinter without having taken a blood meal. They emerge early in the spring, sometimes before the snow has melted away, and search for a warm blooded vertebrate. This species is found throughout western America and is one of the largest in Alaska, often referred to as “Snow mosquito”.

Crane Flies
Family Tipulidae

Adult: length to 1". In this family, we find the largest, long-legged mosquito-like flies. They are common in damp locations and dense vegetation. Some aquatic larvae are predaceous on other invertebrates. The larvae have a tough outer skin and feed on decaying vegetation in moist, damp or water-saturated soil or are active predators in the gravel base of fast-water streams. Adults are short-lived and do not feed.

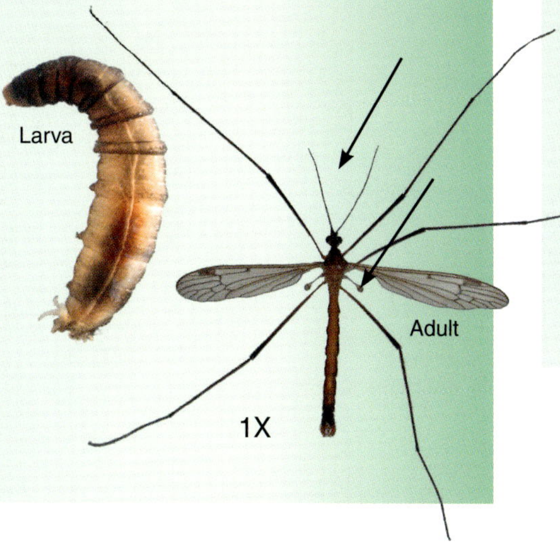

Fungus Gnats
Family Sciaridae

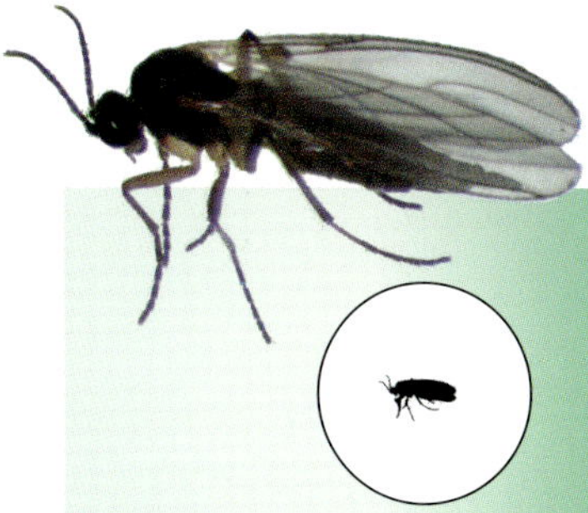

Length: less than 3/8". They are small, long-legged brown flies with dark-colored wings. They are common in greenhouses and flower pots. The larvae live in moist organic soil with some species feeding on plant roots.

Midges
Family Chironomidae

Length: 1/4". Adults are mosquito-like, but do not have a proboscis and do not bite. Males have feathery antennae. Most larvae are gatherer/collectors and live on the muddy bottoms of lakes. Midges are the most abundant insects, in terms of both abundance and number of species in lakes and streams of south-central Alaska.

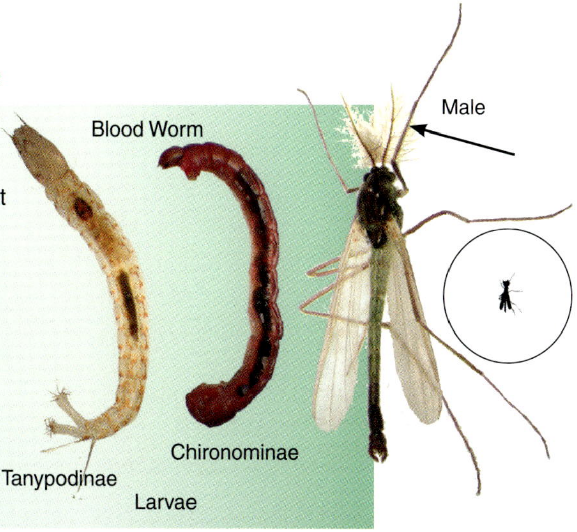

Phantom Midges
Family Chaoboridae

Nymph

These midges resemble mosquitoes but do not have the long proboscis and do not bite. The larva is almost transparent and is found in temporary pools, remaining still and horizontal in mid-water (see p. 66). They are highly predatory on mosquito larvae.

Black Flies, Whitesocks
Family Simuliidae

Length: less than 3/16". The many-segmented antennae of these insects are short and the thorax of the adult forms a large hump. The adults actively crawl around and females bite. The larvae live in fast-moving streams where they filter water for microorganisms. Common species have the lower parts of the legs white contrasting with the black body, hence the name whitesocks. These flies can be extremely annoying on windless days in late summer.

Fungus Gnats
Family Mycetophilidae

Length: up to 1/2". These flies often have a large humped thorax, with the head often not visible from above. Their legs are long and have long stiff hairs at the joints. The wings are often dark colored. Adults are often found on flowers. The larvae feed on decaying vegetation and fungi.

***Mycetophila* sp.**

Gall Midges
Family Cecidomyiidae

***Dasineura* sp.**
Length: up to 1/4". The adults of Cecidomyiidae are weak fliers and are short-lived. The name "gall midges" is somewhat misleading because only some of the species induce galls on plants. Many other Cecidomyiidae are instead predaceous on aphids, feed on fungi or live in the vegetation without inducing galls.

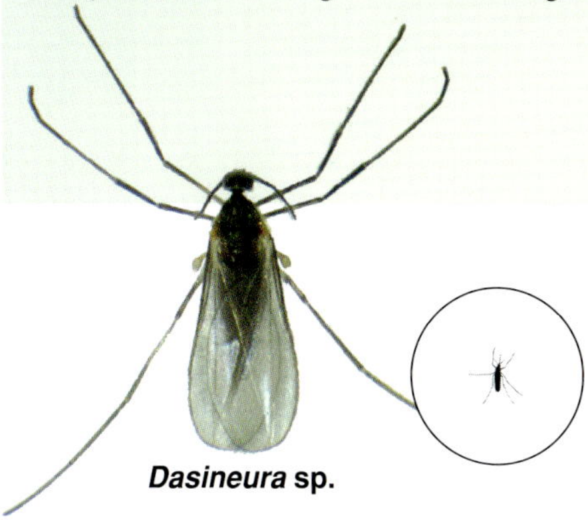

***Dasineura* sp.**

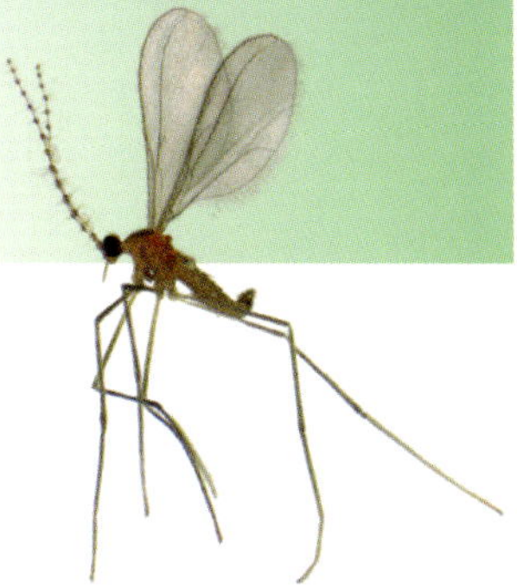

Punkies, No-see-ums
Family Ceratopogonidae

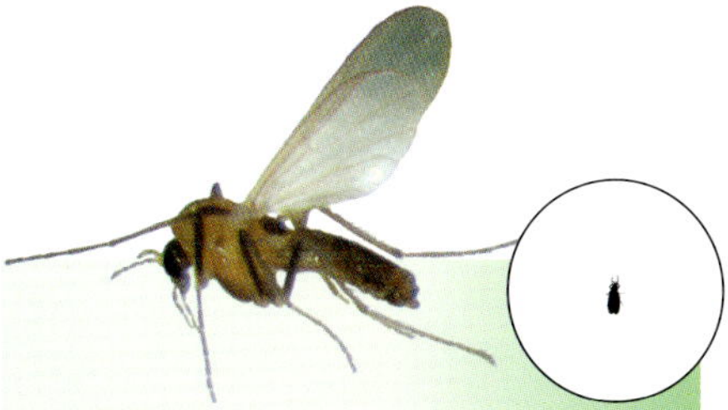

Length: less than 1/16". Minute biting flies. Adults have a large thorax, forming a hump, and moderately long antennae. Only a few species of this family bite people, but these can be very abundant. Biting no-see-ums are small enough to crawl through mosquito screen and can be annoying pests. Most other species feed by biting larger insects. The largest species are predatory on smaller insects. The larvae are long and wiry and they live in wet soils.

Moth Flies, Sand Flies, Bathroom Gnat
Family Psychodidae

***Duckhousiella* sp.**
Length: less than 1/16". These flies look like miniature moths but have a single pair of wings densely covered with long hairs. At rest, the wings are held roof-like on their back. The antennae are long. The larvae of these tiny flies are aquatic or live in moist soil. Adults are common in moist places including outdoor bathrooms. A few species bite.

Botflies and Warble Flies
Family Oestridae

The larvae of these flies are internal parasites of large mammals such as caribou. The adults are fast fliers, resembling bees, but have only one pair of wings and do not have functional mouths.

Caribou Botfly
Cephenemyia trompe

Length: 1/2". The larvae of this large bee-like fly develop inside the nasal cavities of caribou in Alaska.

Late in summer, the female botfly lays live larvae on a caribou's head, near its mouth. The larvae crawl into the mouth and migrate to the back of the throat where they settle. The following spring, the larvae, which are fully grown, detach themselves from the back of the caribou's throat and are coughed to the ground by the caribou. The larvae then penetrate the soil where they pupate and the adult botflies emerge later in the summer. Adult botflies can be encountered in mid summer while hiking the upper forest zone or the subalpine zone of the Kenai mountains.

Male botflies sometimes choose where a seepage crosses a segment of a hiking trail as a territory. Anyone passing through its territory is immediately met by the fly which often executes a few high speed aerial bypass maneuvers for closer inspection before losing interest and returning to his routine pattern of flying from one patch of bare mineral soil to the next in a repeated sequence.

Warble Fly
Hypoderma tarandi

Length: 5/8". Warble flies attach their eggs to the fur of caribou or reindeer and the larvae penetrate under the skin of the back of the caribou where they develop. Early in the spring the larvae emerge from the caribou's skin and drop to the ground where they pupate. Adult warble flies can be noticed flying near mountain ridges in the Kenai mountains. These fast flyers often come close to hikers, checking whether they are caribou or not.

Soldier Flies
Family Stratyomyiidae

Beris fuscipes
Length: 3/8". Among all of the flies with a green metallic thorax, species of the genus *Beris* are unique, because of the finger-like projections on the rear of the thorax. The larvae live in rich organic soils and sometimes emerge from flower pots kept indoors.

Sargus decorus
Length: 1/4". The adults visit flowers and willow catkins. The larvae usually develop in cow dung, and possibly in moose droppings.

March Flies
Family Bibionidae

***Bibio* sp.**
Length: 1/4". These are robust hairy black flies and the males fly in dense swarms that can resemble smoke from a distance. The eyes of males and females differ greatly. Larvae live in decaying wood. Bibionids were the most abundant insects during the middle and late Tertiary period.

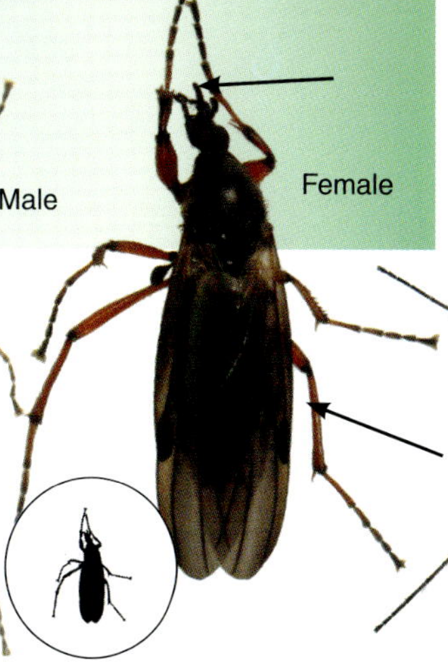

Hesperinus brevifrons
Length: 1/4". These are black flies with long segmented antennae, and their wings are darkened with a black patch toward the tip of each wing. Unlike most other bibionids, *Hesperinus* has long antennae. The larvae are scavengers in decaying organic matter.

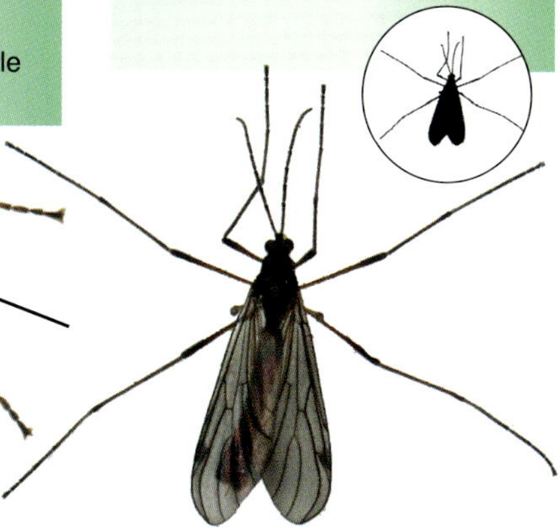

Flower Flies
Family Syrphidae

Most of the flies that hover like hummingbirds near flowers or vegetation belong to this family. In south-central Alaska, the only other insects that hover are hawk moths and big-headed flies. All of these powerful fliers are often seen at flowers, refueling on sweet nectar to replenish their energy supplies. Many of the Syrphinae have their abdomen transversely barred with yellow and black like wasps, while Eristalinae can be robust and fuzzy like bumblebees or look like non-descript flies. The larvae of Syrphinae are voracious predators of aphids while the larvae of Eristalinae are semi-aquatic and live in organic-laden pools, where they feed on microorganisms. Forty-three species of Syrphidae have been identified in south-central Alaska.

Subfamily Eristalinae

These flower flies are covered with dense hair and mimic bees or bumblebees, but they do not have stingers.

Eristalis spp.

Length: 1/2". *Eristalis* aquatic larvae live in water that is rich in organic matter or organic-rich wetlands where they feed on bacteria. They are called "rat-tailed maggots" as their abdomen extends into a tail-like breathing tube. The presence of numerous *Eristalis* larvae is a sure indication that the body of water is full of organic matter.

The male *Eristalis* defends his territory by aggressively chasing anything that resembles a competitor, and actively pursues females that venture into his airspace.

Subfamily Syrphinae

Larvae of this subfamily raid aphid colonies. Lumbering among the crowded aphids, syrphid larvae grab aphids with their powerful mandibles and sling them into the air without letting go. They then suck the aphid dry, discard the empty body shell and immediately grab another.

Larva

female

male

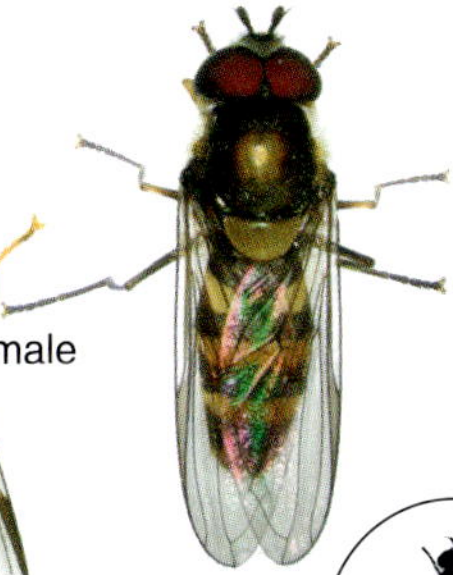

pupa

egg (not shown)

Larva

Pupa

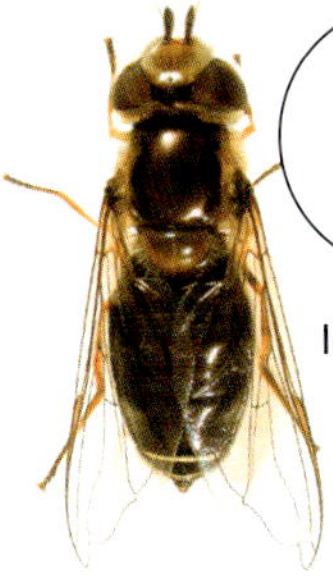

Imago

Life stages of ***Chrysotoxum***
The larva is a voracious aphid predator.

***Chrysotoxum* sp.**
Length: 1/2". The antennae of these insects are long for a flower fly. The yellow and black pattern on the abdomen mimics paper wasps. The maggot-like larvae of these genera are voracious predators of aphids and other Homoptera.

Baccha elongata
Length: 1/2". The compound eyes on males of this insect envelope almost the entire head. The *Baccha* abdomen is extremely narrow at the base and club-shaped at the end. The larvae are aphid predators. This species is found throughout the northern hemisphere.

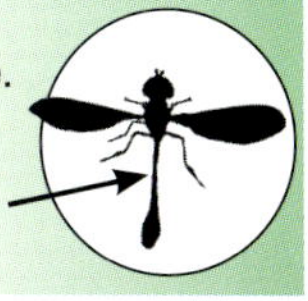

Family Piophilidae

Amphipogon hyperboreus
Length 1/4". Males have strange combs of long hairs on the head, legs and the end of their abdomen. The larvae live in decaying fungi.

Horse and Deer Flies
Family Tabanidae

Twenty-five species in three genera have been identified in south-central Alaska.

Horse Fly
***Hybomitra* sp.**
Length: 3/4". These powerful fliers get most of their energy by lapping honeydew from aphids, but the females of this large fly also need a blood meal to mature their eggs. They attack large mammals, including people. The bite is quite painful and animals often panic when harassed by horse flies. Horse flies are vectors of blood parasites in cattle. Since they are often interrupted during their feeding, horse flies will often attack several different individuals, thus increasing the chance for contamination. The larvae are aquatic and feed on soft-bodied insects. Horse fly larvae are attacked by numerous parasitoids.

Deer Fly
Chrysops excitans
Length: 3/8". These smaller flies have gold and green metallic eyes and dark patterns on the wings. They bite people and are quite annoying. The larvae live in fresh water.

Bottle Flies, Blow Flies
Family Calliphoridae

Length: 3/8". Abdomen and thorax of adult flies are either metallic green, blue or metallic black. They can be difficult to identify to the species. Maggots, (the larvae), develop on carrion and manure. Some species may infect dirty wounds of wildlife or neglected cattle, and thereby cause premature death from bacterial infection. These insects grow very fast and may have half a dozen generations per year.

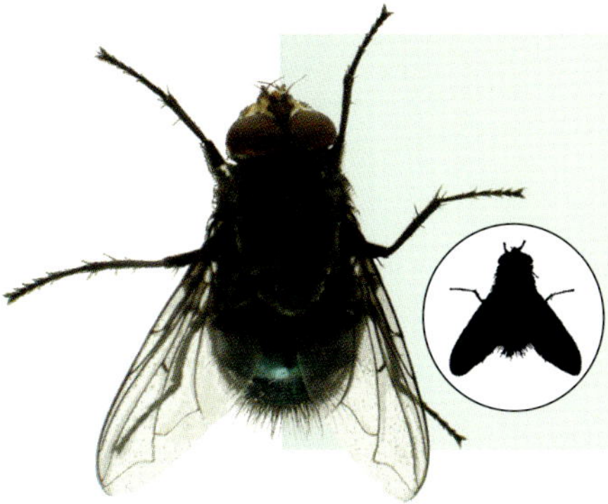

Protophormia terranovae

These very common flies are quick to locate fresh corpses on which they lay their eggs. Forensic entomologists estimate the time of death by examining the stage of development of the maggot or the pupa found on a corpse.

Flesh Flies
Family Sarcophagidae

Length: 3/8". Blackish flies that are difficult to identify the species. The larvae live on carrion and decaying organic matter.

Muscid Flies
Family Muscidae

House fly
Mesembrina solitaria

Length 5/8". The larvae, a maggot, lives in decaying organic matter, including garbage. Very hairy flies, usually black.

Big-headed Flies
Family Pipunculidae

These flies have an elongated black body and compound eyes that almost entirely cover the semi spherical head. Like flower flies (*Syrphidae*), big-headed flies can hover like humming birds. Pipunculidae are endoparasites of leafhoppers. While remaining airborne, the females are able to inject an egg between segments of the abdomen on their host. The larva develops inside the leafhopper until mature, filling the entire cavity of its host's body. The parasitoid larva then emerges and pupates in the leaf litter. Big headed flies are important in the biological control of agricultural pests.

***Dorylomorpha* sp.**
Length: 1/4". The female of this species lays one egg inside the body of its host, the hopper. It can parasitize 2-3 hoppers a day and can live as an adult for four days.

Robber Flies
Family Asilidae

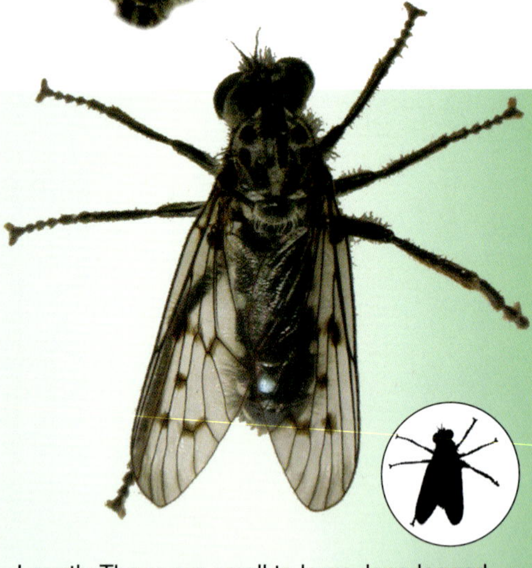

Length: These are small to large, long-legged, hairy flies. The upper part of the head in between the eyes is concave. Adults are predators. The larvae live in the soil where they eat the eggs, larvae or pupae of other insects.

Snail Killers
Family Sciomyzidae

***Elgiva* sp.**
Length: 1/4". This fly can be found near lakes and ponds. The larvae are predators of snails and freshwater clams. Species of this genus are found throughout the northern hemisphere.

Tachinid Flies
Family Tachinidae

These are small to medium-sized, densely hairy flies. Adult tachinid flies resemble common house flies but the larvae have a very different biology. While house fly maggots feed on decaying organic matter, larvae of tachinids are internal parasitoids of many insects, especially Lepidoptera.

The female tachinid deposits its egg on the host, injects it inside the host or disperses minute eggs near the host, which are then ingested by the host. The tachinid larva develops inside the host and emerges as an adult usually after the host dies.

Tachinid flies are difficult to distinguish from an ordinary fly. The best way to find tachinids is to rear parasitized caterpillars. Caterpillars parasited by tachinids can be identified by the tachinid larva's breathing openings found on the back of the caterpillar. Later in development, the body of the caterpillar will become distorted as the parasitoid develops inside the host.

Tachinidae have an important role in controlling the population of many moths and other common insects. The parasitism rate is often 50% and sometimes close to 100%. Tachinid flies have been used successfully to control agricultural pests in other regions.

Breathing opening of the tachinid larva inside its host.

Caterpillar parasited by a tachinid fly

Family Odiniidae

These flies are captured mostly in the vicinity of dead trees attacked by boring insects. The family is comprised of four genera and eleven species in North America.

Length: 1/8". The larvae live inside the galleries of wood-boring insects. It is not clear if the larvae feed on fungi or if they are parasitic on the wood-boring insect larvae.

Humpback Flies
Family Phoridae

Length: less than 1/4". Adults are small, usually black, sometimes light colored. Their hind legs are usually very flat, the thorax has a large hump, the head is small, with long hair and very short antennae. Larvae either live in decaying vegetation, are parasitic on other insects or live inside ant nests.

Lance Flies
Family Lonchaeidae

Lonchaea **sp.**
Length: 1/4". The body of these insects is shiny black and hairy. The wings are clear and yellowish at the base. The larvae are either predators of bark beetle larvae and pupae or feed on decaying frass under the bark.

Louse Flies
Family Hippoboscidae

Head, side view

***Ornithomya* sp.**

Length: 3/16". The body and head of these insects is strongly flattened. This unusual fly is an external bird parasite. It lives under the feathers of wild birds, but is able to fly on its own. The sheep ked is a related wingless species that causes harm to domestic sheep, especially young lambs, if left unchecked. Female Hippoboscidae give birth to only a few (7-9) larvae during the course of their lifetime.

Xylophagid Flies
Family Xylophagidae

***Xylophagus* sp.**

Length: 1/2". The elongated body of these insects is covered with few hairs and their wings are tinted dark. They have a black mark on their fore edge and their abdomen is tapered to a point. Larvae live under the bark of dead trees or in decaying wood. The flies are mostly found in forested habitats.

Long-legged Flies
Family Dolichopodidae

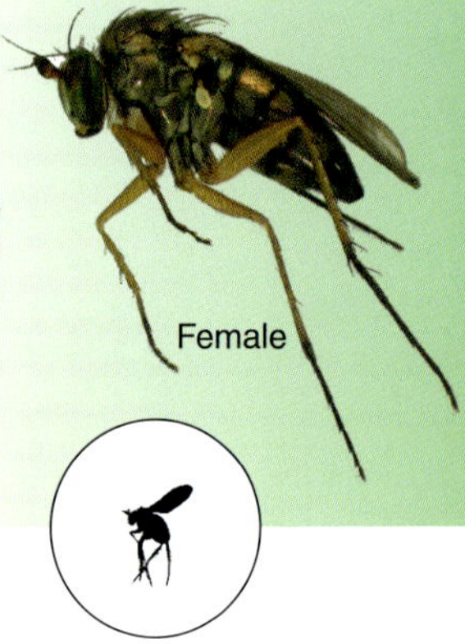

***Dolichopus* spp.**
Length: 3/16". Adults are metallic green or coppery. The genitalia of the male is often folded under the abdomen while females have a straight, pointed abdomen. These insects are very common, especially in moist habitats. The larvae live in decaying vegetation, under tree bark or in wet habitats. The adults are predaceous on other small insects.

Dance Flies
Family Empididae

***Rhamphomyia* sp.**
These flies are usually black, but sometimes they are yellow. The females have a conical, pointed abdomen and males have a broad appendix at the end of the abdomen. Both adults and larvae are predators. Males often form dense swarms that attract females. Species from this family can be identified by the short proboscis and their general shape.

MOTHS AND BUTTERFLIES ORDER LEPIDOPTERA

Butterfly and moth wings are covered with minute scales that rub off when handled. The larvae are caterpillars and can be distinguished from sawfly larvae by the presence of several ocelli on each side of the head, and up to 5 pairs of false legs. Sawfly (Hymenoptera, see p.161, 162) larvae have a single eye on each side of the head, and more than 5 pairs of false legs that are not armed with microscopic hooks.

The caterpillars pupate prior to changing into butterflies or moths. Adults feed on nectar from flowers. Most butterflies fly by day and can be easily identified by sight to the genus or to the species level. Moths usually fly at night, few of them are particularly distinctive. Thirty-two families of Lepidopterans have been identified in Alaska.

Skippers Family Hesperidae

3X

Arctic Skipper
Carterocephalus palaemon

Wingspan: 1-1/4". The adults of this species are small, stocky butterflies, with curved antennae tips and erratic flight.
The caterpillar feeds on grasses. This species is found in moist woodlands and in subalpine meadows.

Gossamer-winged Butterflies Family Lycaenidae

3X

Spring Azure
Celestrina lucia

Wingspan: 1". The adults have bluish metallic wings. The caterpillar possibly feeds on Labrador tea. Some species whose caterpillars feed on flowers have a mutualistic relationship with some ants. The caterpillars secrete a sweetened liquid used by the ants, and the ants protect the caterpillars from predators. In Alaska several species similar to *Celestrina lucia* exist but are difficult to distinguish.

Brush-footed Butterflies
Family Nymphalidae

These butterflies do not use their reduced forelegs and appear to have only four legs. Most of the butterflies in this family are robust, living several months. They have colorful patterns on the upper side of the wings and are camouflaged on the undersides, allowing them to blend in with the background when they are at rest with their wings closed.

Mourning Cloak
Nymphalis antiopa

Wingspan: 2-1/4" to 3". This is a very common butterfly with dark purplish-brown wings bordered with a yellow band. The adults overwinter and are some of the first butterflies to fly in the spring.

The caterpillars, spiky with black and red spots, feed on willow and cottonwood in groups when young, becoming solitary later on. The mourning cloak is widespread throughout the northern hemisphere.

1X

1X

Green Comma, Anglewing

Polygonia faunus

Wingspan: 1-1/2". The wing margins of this butterfly are deeply notched, and appear torn up. The common name refers to the comma-shaped spots on the dark underside of the wings. Caterpillars are yellow-orange with branched spikes, and they mostly feed on willows and birch. This species is widespread throughout North America.

1.5X

Tortoise Shell

Aglais milberti

Wingspan: 1-3/4".

The caterpillar feeds on nettle leaves. Tortoise shells are widespread throughout the northern hemisphere.

Swallowtails and Parnassian
Family Papilionidae

Large colorful butterflies with all six legs functional.

Tiger Swallowtail
Papilio canadensis

Wingspan: 3". This is the largest butterfly of south-central Alaska. The hind-wings are marked with several orange spots. The caterpillar is marked with false eyes. This species is common in disturbed sites of the forested zone where food plants such as Barclay willow, cottonwoods and aspens grow. It is widespread throughout North America.

Old World Swallowtail
Papilio machaon

Wingspan: 2-1/2" to 2-1/2".
This butterfly can be distinguished from the tiger swallowtail by the single orange spot on the hind wing. The caterpillar feeds on sage and wormwood. In south-central Alaska, it is found in subalpine meadows near or above treeline. This species is found in northern Canada, Alaska, northern Asia and Europe.

Phoebus Parnassian, Apollo
Parnassius phoebus
Wingspan: 2" to 2-1/2". This is a large white butterfly with black markings and red spots circled with black. The caterpillar is black with red spots and feeds on saxifrages. Parnassians are found only in the alpine zones. It is widespread throughout the mountains of western North America.

1X

Whites and Sulfurs
Family Pieridae

Mustard White
Pieris angelika
Wingspan: 1-3/4". The wings of this butterfly are entirely white with grey veins. It is found around roadsides, subalpine meadows and on disturbed sites in damp forests. The caterpillar is green with a white stripe and it feeds on mustard plants. This species is found in northwestern America. It is closely related to *Pieris napi.*

2X

Clouded Sulfur
Colias philodice
Wingspan 1-1/2". The wings are bright yellow and the spot on the hindwings is orange. The caterpillar is green with a dark dorsal stripe and pink lateral line. It feeds on legumes, especially clover.

1.5X

Tiger Moths or Woolly Bear Caterpillars
Family Arctiidae

The moths often have bright color patterns on the wings. Some of these moths fly by day, others are nocturnal. At rest, the moths fold their wings roof-like over their back.

These caterpillars are densely hairy caterpillars, commonly called "woolly bear caterpillars". They are often noticed in early spring, even before all the snowpack has melted away, searching for a suitable site to pupate. Their long hairs, which are an irritant and deter predators, are incorporated into the fabric of the cocoon.

"Woolly bear" tiger moth caterpillars curl up when disturbed.

Wood Tiger Moth
Parasemia plantaginis
Wingspan: 1-1/4" to 1-1/2". This moth is common in open woodlands. The male flies during daylight while the female is more nocturnal. The caterpillars feed on a variety of plants including birch, blueberry and plantain. This moth is commonly seen in June in the upper forests and subalpine zones and is found in the western half of North America.

Male

2X

Female at rest

Ruby Tiger Moth
Phragmatobia fuliginosa
Wingspan: 1-1/4 ". The wings of this moth are dark red and the body is densely hairy. The caterpillars feed on dandelion. This moth is often parasitized by braconid wasps. This species is found throughout the northern half of North America.

2X

Arctia opulenta
Wingspan: 2-1/2". The forewings are brown with a pattern of white bands, the hindwings are orange with black spots. The caterpillar is a generalist feeder. The species is found throughout North America.

1.5X

1.5X

St. Laurence Tiger Moth
Platarctia parthenos
Wingspan: 2-1/2". This moth flies during daylight hours in early summer. The caterpillars are generalist feeders. This species is widespread in western North America, but is not abundant.

1X

Clearwing Moth
Family Sesiidae

These moths have large areas of their wings transparent because they lack the scales that cover the entire wings of most moths. Because of the long transparent wings, adults of sesiid adults resemble wasps more than moths. Their flight is strong and direct. The larvae of the clearwing moth bore into wounded birch.

Albuna pyramidalis
Length: 1/2" to 3/4". This species is common, while 2 other species of *Albuna* are rare in Alaska. The larvae feed on fireweed. *Albuna pyramidalis* is found throughout North America.

Hook-tip Moths
Family Drepanidae

The most obvious characteristic of these small moths is the tip of their front wings which extend into hooks. The adults fly at night. Only one species has been reported in Alaska.

Warty Birch Caterpillar
Drepana bilineata
Wingspan: 1-1/2". The forewing of the adult moth is pale brown covered with fine transverse brown lines. This species is uncommon. The caterpillar feeds on paper birch alder and poplar. This species is found throughout western North America.

Loopers
Family Geometridae

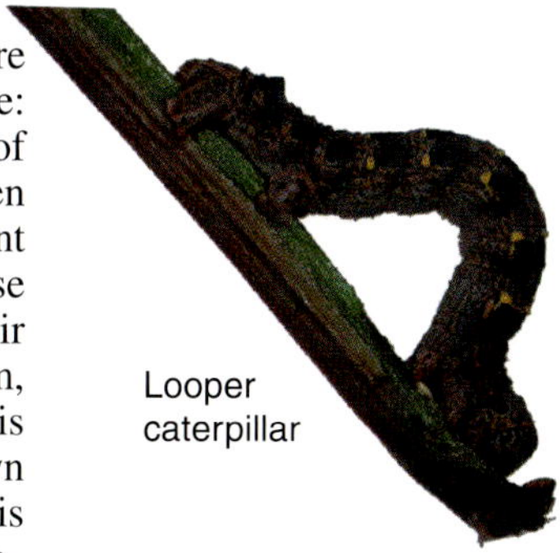

Looper caterpillar

Many of these moths are fairly delicate and are weak fliers. The caterpillars are very distinctive: they have functional legs only at the extremities of their long body. To travel, they arch their abdomen and position the rear section right behind the front section of legs. When the rear is secure, they release the front legs, stretch the anterior section of their body forward, grip the substrate in front of them, release the rear section and repeat the motion. This is compared to the action of a surveyor, also known as geometer, measuring a distance longer than his measuring tape, hence the family name Geometridae.

Most caterpillars of these moths are cryptic. That is, they look like twigs or conifer needles and are difficult to detect. The best method to collect them is to lay a white tarp under a tree or a shrub and beat on the branches with a stick to dislodge the caterpillars.

Pale Beauty, Fringe Looper
Campaea perlata

Wingspan: 1-1/2". This is a common, delicate moth. The caterpillars measure up to 1-1/2", are smooth above and have a long, low fringe of white hair low on either side of the body. These pale hairs diffuse the shadow of the caterpillar, making it harder to detect. The caterpillars feed on a variety of shrubs and trees such as alder, birch, willow and aspen. This species is found throughout North America.

Two-pronged Looper, Sharp-lined Yellow
Sicya macularia

Wingspan: 1-1/4". This is a delicate, yellow moth with an oblique line across the forewings. The caterpillars measure up to 1" long, have 2 distinct, vertical finger-like lobes rising from the middle of the body.

The caterpillars eat the leaves of a variety of shrubs and trees such as alder, aspen, blueberry, currant, poplar and willow.

This species is uncommon. It is found throughout North America.

1.5X

Spear-marked Black Moth
Rheumaptera hastata

Wingspan 1" to 1-1/4". Black and white moths, usually with a "spear" mark on the fore and hindwings. The black spear-marked moths appear in early June or early July and live 3 weeks. Females attract males with pheromones, mate and lay eggs. After emerging, the young caterpillars are gregarious and find shelter between two leaves sandwiched together with silk. Later they disperse and each moves into its own sandwich leaf shelter. These caterpillars feed by removing the upper layer of the leaves. When fully grown they lower themselves to the ground on silken threads. The spear-marked moths overwinter as pupae inside a cocoon in the leaf litter. Populations are usually kept in check by an assortment of parasitoids, predators such as redpolls and chickadees, and diseases. Survival of the pupae is compromised by low temperatures with light snow cover. In Interior Alaska, massive outbreaks occur every 12 to 15 years. Millions of acres of birch forest can be affected. After an outbreak, the moth population naturally declines within 2 years. Common in south-central Alaska and throughout North America.

Male

Moth at rest

1X

2X

Male

Female

Leucobrephos brephoides

Wingspan: 1-1/4". The adult body is very hairy, grey with white patterns. This species is found in the boreal forest throughout North America. The caterpillars, green with yellow stripes, feed on willow, poplars, birch and alder.

Cladara limitaria

Wingspan 1" to 1-1/4". This drab-colored moth is frail, and flies early in the spring. The caterpillar has the particularity of resting with its head tucked under its thorax. These green caterpillars feed on developing willow, birch and alder leaves and are ready to pupate by June. The moth overwinters as a pupa. *Cladara limitara* is found throughout northern forested regions of North America.

2X

Leaf Blotch Miners Family Gracilariidae

The line of feces in the blotch or mine identifies the small moth as *Phyllocnistis populiella.*

Lithocolletis

The larvae of the leaf blotch miners tunnel between the two outer layers of aspen leaves. The flat caterpillar pupates at the end of the tunnel.

Pyralid Moths Family Pyralidae

The wings of Pyralidae are quite variable: for some species, the wings are narrow, while for others the wings are broad and triangular. Most pyralid moths have mouth parts that extend into a snout. Pyralid larvae bore into the phloem, and into spruce cones, feed on foliage or are even predators on insects. The Indian meal moth, *Plodia interpunctella* (not illustrated) is a common household pest.

2X

***Udea* sp.**
Wingspan: 1". The resting moth is triangular in outline. The green caterpillar has yellow longitudinal lines on the body. These larvae are found both outdoors and in greenhouses.

***Crambus* sp.**
Wingspan 1". The adult moths of this species hold their wings folded along their bodies while at rest. The caterpillars feed on lawn grasses and spin a silky shelter among the vegetation.

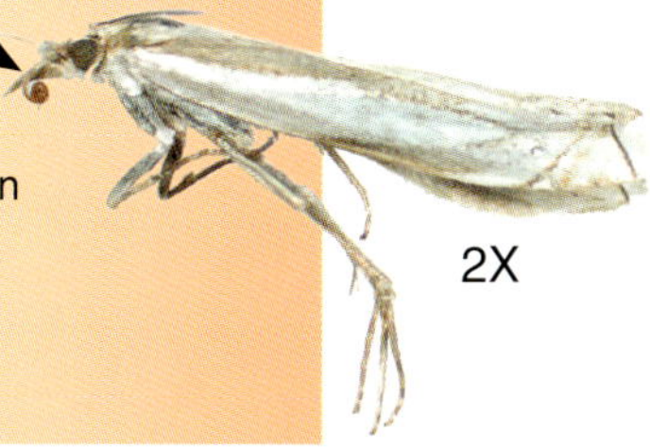

2X

Torticids
Family Tortricidae

These moths are drab-colored, their forewings squared at the tips. The caterpillars are leaf tiers or leaf rollers. They pupate in rolled leaves or in bark crevices. Some species are agricultural pests. There are at least twenty-nine genera and one hundred and ten species in this family in Alaska.

Large Aspen Tortrix
Archips fractivittana

2X

Wingspan: 1". The grey-green caterpillars with black heads emerge in August and feed on aspen trees, then overwinter in a sheltered site. In the spring, the larvae become active and hollow the winter buds, which can cause the complete destruction of the young leaves. A population can reach outbreak level over large areas and completely defoliate all aspens in a region. Archips sp. caterpillars also feed on white, Sitka and Lutz spruce as well as Prunus. Five species of Archips are known to be in Alaska. *A. fumiferana* (spruce budworm) can completely defoliate and subsequently kill spruce trees. Another species, *A. rosaceana,* feeds on mountain ash.

White Triangle Tortrix
Clepsis persicana

Wingspan 5/8" to 3/4". The adult moth has a large white triangular patch on the front edge of the forewing. Caterpillars live on alder, birch and spruce. *Clepsis persicana* is known from most of North America.

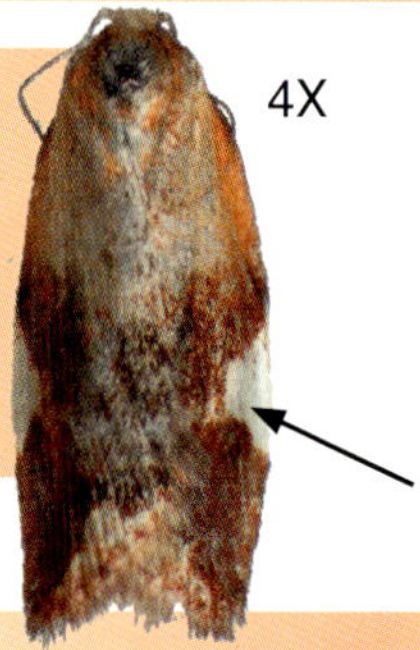

4X

Family Gelechiidae

Wing span: 3/4". The palpi (mouth parts) are large and curved upward. This family includes numerous species of small moths. Larvae are leaf miners, leaf rollers or leaf tiers. At this writing, five genera and six species have been identified from Alaska.

3X

Tussock Moths
Family Lymantriidae

Rusty Tussock Moth
Orgyia antiqua

Wingspan: 1". The male moths, orange to dark brown with a white spot on the forewings, fly erratically during the daylight hours. The females look very different, and are unable to fly because they lack wings. Their bulging abdomen is covered with dense grey hairs. As soon as the female emerges from her cocoon, it attracts males with pheromones, mates and lays its eggs on the webbing surrounding her now-empty cocoon. The young caterpillars disperse on a windy day by hanging on a thread of silk they secrete. This showy caterpillar, covered with long hairs and four dense tufts of black hairs on the back, measures up to 1 1/4". They are eclectic in their choice of food plants, eating the leaves of at least eight species of willows, in addition to numerous other shrubs and trees such as birch, cotoneaster, blueberry, birch and white spruce.

Caterpillar

Flightless female

Male color variations

Eggs on female's empty cocoon

All 1.5X

Prominent
Family Notodontidae

Furcula occidentalis

Wingspan: to 1-1/4". The moth is pale grey, and has elongated wings.

The caterpillars have a very distinctive yellow green body with a broad dark brown saddle, and feed on balsam poplar and Barclay willow.

This moth is found occasionally in Alaska. It is more abundant in eastern and southwestern states

Male moth

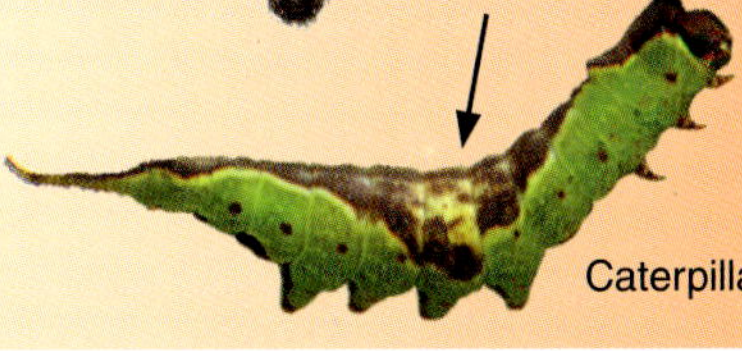

Caterpillar

Cutworms
Family Noctuidae

The moths of this family are mostly cryptic like *Autographa,* which is almost invisible when resting on birch bark. A few noctuids like *Androloma* and *Alypia* are brightly contrasted. The caterpillars of this family are hairless and most remain hidden during daylight hours. Some live at the base of plants where they eat the roots. Several noctuid larvae, also known as cutworms, cause damage in gardens. This is a large family that includes twenty-seven hundred species found though out North America.

Enargia infumata
Wingspan: 1-1/4". The caterpillar lives on poplar, birch and willows. This species is found throughout North America.

2X

2X

Scoliopterix libatrix
Wing span: 1-3/4". The outline of the forewings of this moth is irregular, and there is some orange-red color near the body. The moth overwinters in the adult stage. The caterpillars, velvety green and measuring up to 2" long, eat the foliage of willow and poplars.

2X

Semilooper
***Autographa* sp.**
Wingspan: 1-5/8". This moth is mottled grey or brown with a white or yellow comma-shaped spot in the middle of its forewing. The moth is well camouflaged when resting on birch bark with lichens. Newly emerged moths have tufts of hair on the back that break up its silhouette and help it resemble the rough bark scars on trunks. The greenish caterpillars feed on the leaves of various plants. Several species resemble *A. buraetica.* Semiloopers can also be pests in cultivated gardens and greenhouses.

Anarta melanopa
Wingspan: 1". Mountain species, found throughout the northern hemisphere.

***Lasionycta* sp.**
Wingspan: 1-1/4". Alpine species. Possibly distributed throughout the Northern hemisphere

During mid summer, the larva of *Lacinipolia olivacea,* up to 1" long, burrows right beneath the surface of sand dunes among beach peas. The raised tunnel can be traced over several feet as the larvae travel to feed from one shoot of beach pea to another.

1.5X

Androloma maccullochi
Wingspan: 1-1/2". Compare the wing pattern with *Alypyia langtoni* and *Parasemia plantaginis.* (See p. 148). *A. macculochii* is widely distributed in North America.

1X

Langton Forester
Alypia langtoni
Wingspan: 1-1/4". The adults of this species are diurnal. The forewings have two round cream-colored spots and the hind legs contain some orange tufts of hair. These hairy caterpillars feed on fireweed. *Alypia langtoni* resembles *Androloma maccullochi,* which can be distinguished by its forewings which are black with broad white spots. This moth sometimes reaches outbreak levels. This species is found in the northern half of North America.

1.5X

Parastichis suspecta
Wingspan: 1-1/4". This caterpillar feeds on birch and willow and is found in the northern half of North America.

1.5X

Polychrysia esmeralda
Wingspan: 1-1/2". The wings of this species have a delicate silky shine and metallic gold color. The caterpillar feeds on Delphinium. This species is found in western Canada, Alaska and the Asian Far East.

1.5X

Eurois occulta
Wingspan: 2". This caterpillar feeds on woody plants and is found throughout most of North America.

Hawk Moths
Family Sphingidae

The hawk moths fly like hummingbirds. Caterpillars, often known as hornworms, have a "horn" at the rear of the abdomen.

Hyles gallii

Wingspan: 2-1/4" to 3". Caterpillars of this species measure up to 2-3/4" long and have a green and a black color phase. They feed on fireweed and river beauty. The number of hawk moths observed each year seems to fluctuate greatly. Recently, high numbers were noticed during the summers of 1980, 1985 and 1993.

Clearwing sphinx moth
Hemaris thysbe

Body length: 3/4". This small hawk moth mimics a hummingbird and feeds in the day by hovering in front of flowers. The caterpillar feeds on shrubs such as high bush cranberry. The species is found throughout North America. It is common in the Matanuska Valley.

WASPS
ORDER HYMENOPTERA

The order includes the wasps, bees, ants, parasitic wasps and sawflies.

Horntails
Family Siricidae

The body of these wasps is cylindrical, and broad at the waist where the thorax meets the abdomen. The females lay their eggs deep into the wood of trees that have been killed or damaged by fire or by other insects. The female inoculates the tree with a fungus. The boring larva, large with a black horn-like protuberance at the end of its tail, feeds on the fungus growing in the galleries. The larvae may take several years to mature. These wasps are impressive with their long ovipositor, but are harmless.

Broad waist ventral view

Xeris
Xeris spectrum

Body length: 1**"**. The ovipositor is long and there is a white spot behind the eye. This species is not common but is distributed throughout the northern hemisphere.

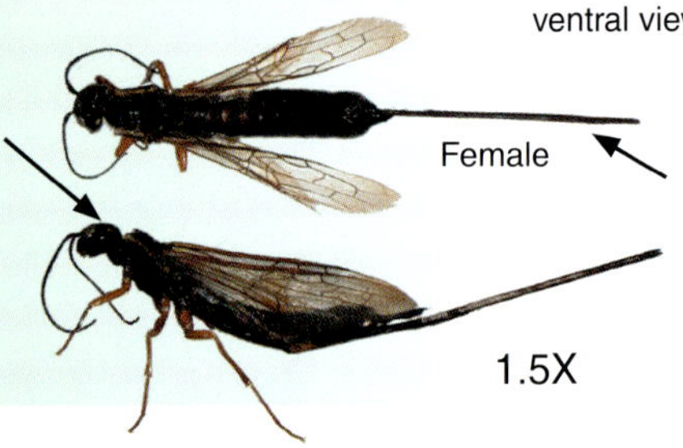

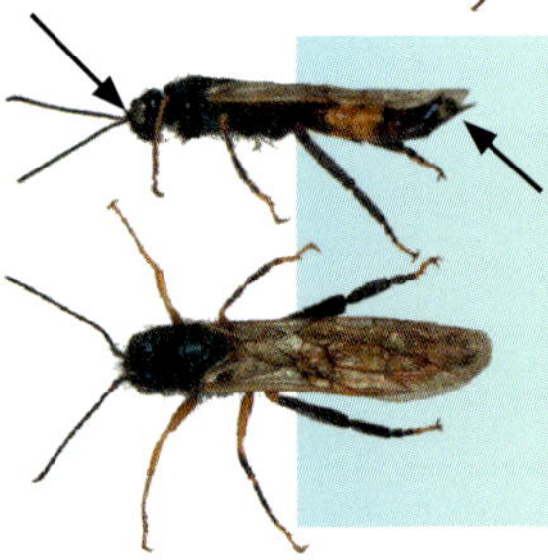

Blue Horntail
Sirex cyaneus

Length 1/2 **"**. The head of this insect is entirely black, the thorax is deeply sculpted and metallic blue. The ovipositor is short. The larvae bore into spruce trees. This species is found throughout the northern hemisphere.

Horntail Wasp
Urocerus gigas

Length: 1**"**. There is a large yellow patch behind the eye of this wasp and the black abdomen is barred with yellow. The larva bores inside white spruce logs and has a sharp hardened point at the end of the abdomen. This large wasp is the most common of this family in Alaska.

Wasps may emerge from fresh house logs. The horntail wasp does not attack dried logs nor does it sting people.

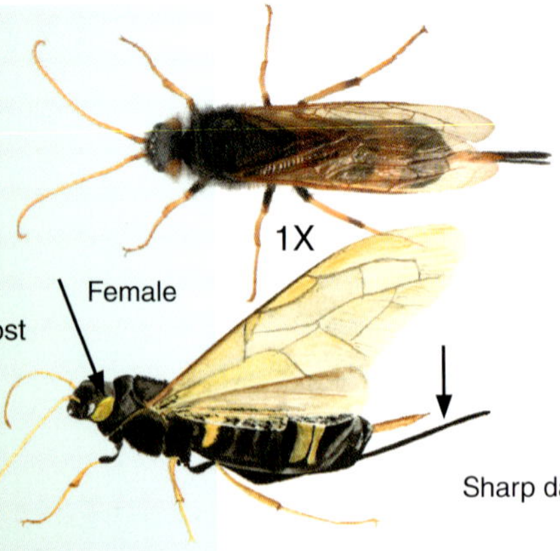

Cimbicid Sawflies
Family Cimbicidae

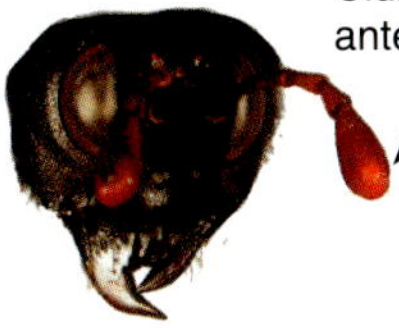

The adults of these wasps are sluggish. Their body is cylindrical, and broad where their thorax meets the abdomen. The antennae are clubbed. The caterpillar-like larvae live on the foliage of broadleaf trees. Typically, they curl up in a tight coil when resting.

Trichiosoma triangulum

Length: 1/2" to 3/4". Hairier and smaller than *Cimbex*. The green larvae live on willow, poplar, aspen, alder and birch. The larvae overwinter inside tough oblong cocoons attached to the twig of a shrub. This species is found throughout North America.

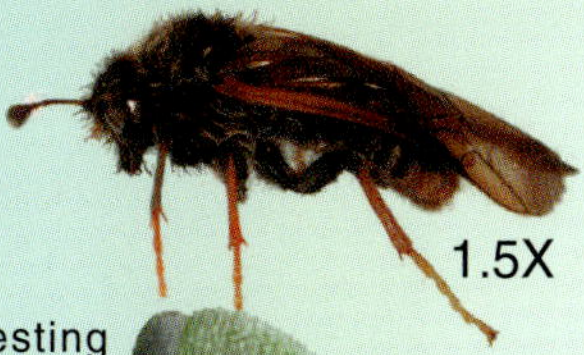

Cimbex americana

Length: up to 1". These large, stout wasps do not have a stinger, but their mandibles are impressive. The male abdomen is black and the female has symmetrical yellow patterns on the abdomen. The larva is yellow to pale green with a black line running down the back. *Cimbex* overwinter as larvae inside cocoons in the top layer of soil. There are several closely related species widespread throughout the northern hemisphere.

Sawflies
Family Tenthredinidae

Species of this family are not well known by the general public. This name is misleading. These insects are not flies (Diptera) but are wasps (Hymenoptera). They have two pairs of wings, but like the two previous families, they have a broad waist. To add to the confusion, the larvae resemble caterpillars. A few sawflies (*Pontania, Euura* etc.) induce galls or mine leaves. Most others are free-living herbivores.

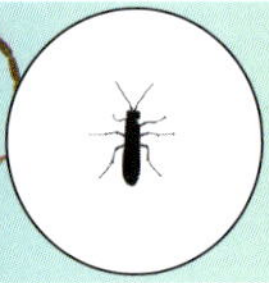

***Euura* sp.**
This genus is very close to *Pontania* but it induces galls on willow stems instead of the leaves, or it may live inside galls induced by other insects on willows.

Larch sawfly
Pristiphora erichsonii
Length: 3/8". Adults have a shiny black body with a broad red band across the abdomen. The larvae feed on larch needles. This species has been introduced in America from Europe. It can completely defoliate Siberian larch grown as an ornamental in south-central Alaska.

Adelesta novae
Body length 3/8".

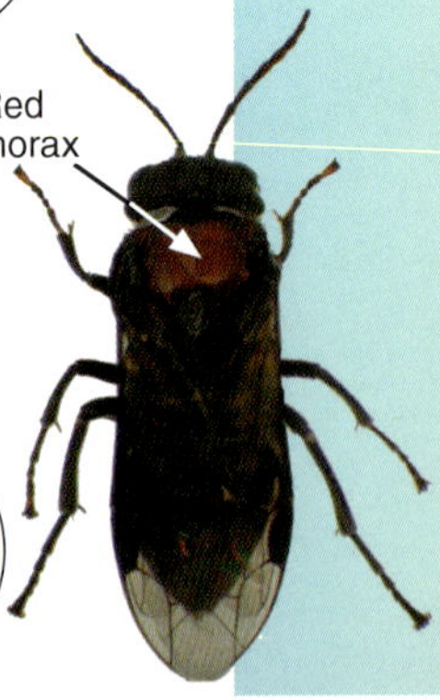

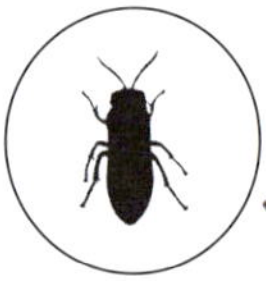

Alder Woolly Sawfly
Eriocampa ovata
Length: 5/16". Adult females are black with large red area on the back of the thorax. The green larvae are covered with fluffy white waxy secretion. This sawfly was introduced from Europe and causes major defoliation of alders in south-central Alaska. Only females are known in North America. This mode of reproduction without mating is called parthenogenesis. The larva is shown on p. 70.

Rhogogaster viridis
Length: 3/8". The adults are green with a black pattern. They are predators on flies that visit flowers. The genus is found throughout the northern hemisphere.

Tenthredo sp.
Length: 1/2" to 5/8". Like *Rhogogaster,* the adults are predatory on insects visiting flowers. They can be observed feeding on nectar of cow parsnip. Occasionally, they also pounce on diptera in an attempt to feed on them. The genus is found throughout the northern hemisphere.

Tenthredo sp.

Amauronematus sp.
Length: 3/8". The larvae feed on willow catkins. The genus is found throughout the northern hemisphere.

Family Argidae

Argidae are distinguished from other sawflies by the antenna with the segments, the last one the being the longest.

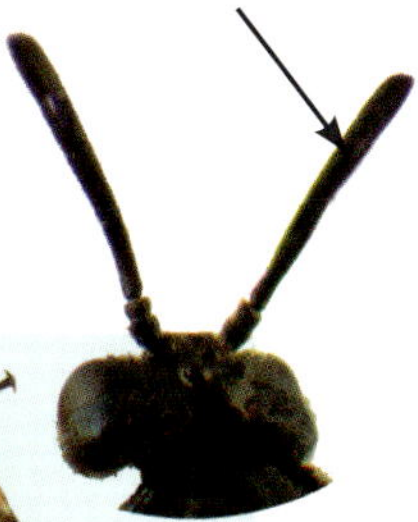

Arge sp.
Length: 1/2". The adult tends to have the wings half-closed on the back when at rest. Larvae are generalist browsers. The genus is known from most continents except Australia.

Braconid Wasps
Family Braconidae

Species in this very large family of wasps are either internal or external parasitoids of insects. They are usually primary parasitoids and kill their hosts, which are Lepidoptera larvae, sawflies and Coleoptera. Braconidae pupate in a spun cocoon outside of the host. Braconidae are closely related to Ichneumonidae, but are usually less than 1/2" long, their female ovipositor is shorter, and most are not as narrow at the waist as ichneumonidae. They can only be differentiated with confidence by comparing the pattern of the veins in the wings. Braconidae are important for controlling many herbivorous insect species.

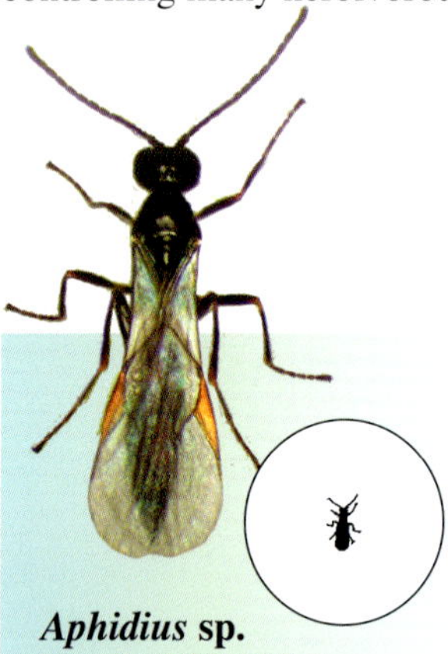

***Aphidius* sp.**
Length: 1/16". This is one of the numerous species that exclusively parasitizes aphids. They are found throughout the world.

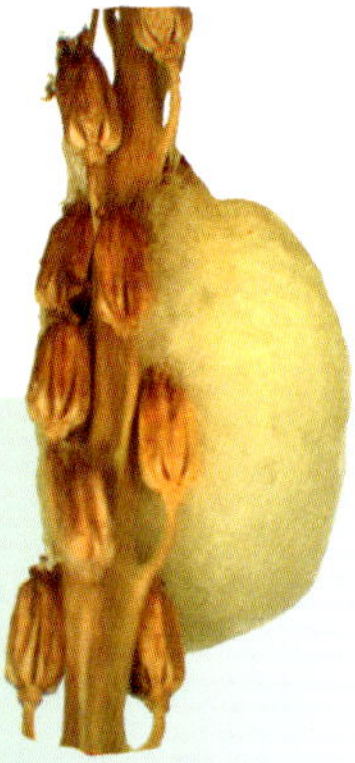

***Cotesia* sp.**
Length: 1/8". After emerging from a single caterpillar, 50 or 60 of these Braconidae spin a communal woolly yellow cocoon 3/4" long and 1/2" wide attached to the top of a stem. They pupate inside the cocoon. The cocoon depicted above was found on the top of an arrow grass stem.

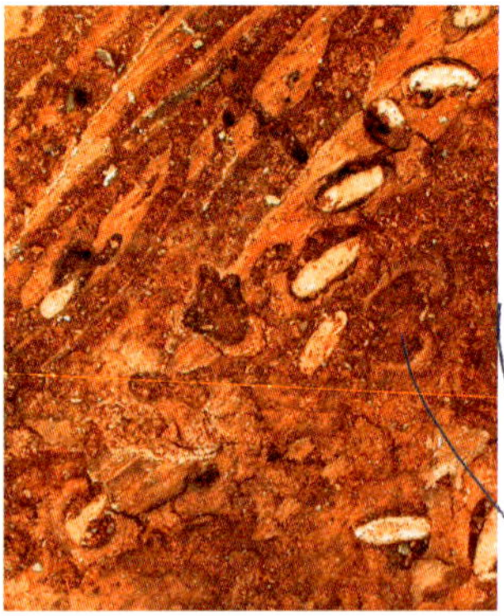

Each cocoon of *Coeloides dendroctoni* found in the gallery of the spruce bark represents one bark beetle that will not contribute progeny to further the outbreak.

Coeloides dendroctoni
Length: 1/4". The larva is ectoparasitic on spruce bark beetle larvae. With its long ovipositor, the parasitoid female oviposits through the bark onto a fully grown scolytid larva. After the larvae are fully grown, they pupate inside the gallery and emerge through circular holes bored through the bark.

Ichneumons
Family Ichneumonidae

This family is the largest of Hymenoptera and includes over 60,000 species worldwide. Ichneumons are differentiated from Braconidae by vein patterns in the forewing, but are usually also distinctive by their larger size, very narrow waist, and brighter color. Adult females often have a very long ovipositor. Ichneumonidae are internal parasitoids of moths and butterflies, beetles, wasps, spiders and diptera. Most ichneumonids develop entirely inside their host and emerge as adult.

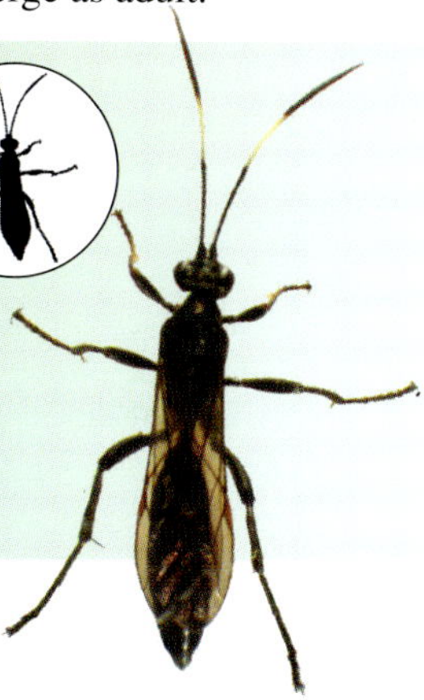

Subfamily Ichneumoninae
Length: 5/8". This wasp was reared from a cocoon found deep inside a rotten birch log. They parasitize both the caterpillar stage and the pupae but always emerge from the pupa.

Subfamily Banchinae
Length: 3/8". Wasps of this subfamily are Lepidopteran internal parasites.

***Ichneumon* sp.**
Length 1/2". These very common wasps are endoparasitods of noctuid moths.

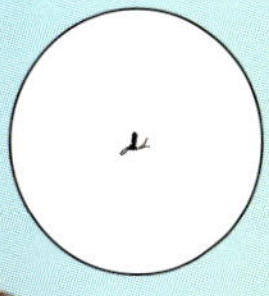

Subfamily Ctenopelmatinae
Lathrolestes luteolator
Length: less than 1/16" . This species has been introduced to the Anchorage, Alaska area in order to control the amber-marked birch leaf miner *(Profenusa thompsoni)* which was accidentally introduced from Europe. The leaf miner sawfly defoliated 30,000 acres of birch in the Anchorage Bowl in 2003.

Subfamily Ryssinae

Larvae are parasitic of wood-boring wasps and beetles.

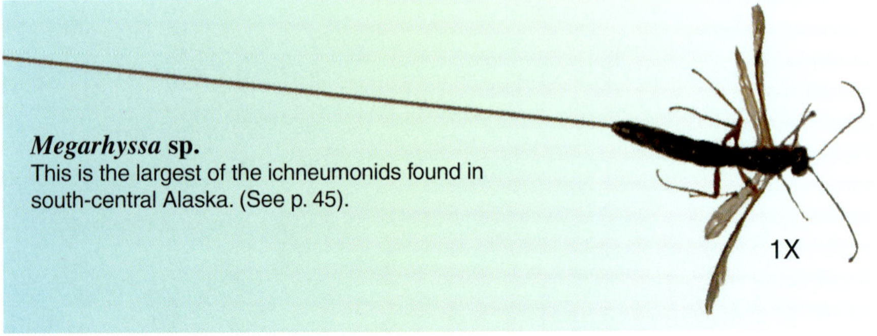

***Megarhyssa* sp.**
This is the largest of the ichneumonids found in south-central Alaska. (See p. 45).

Subfamily Cryptinae

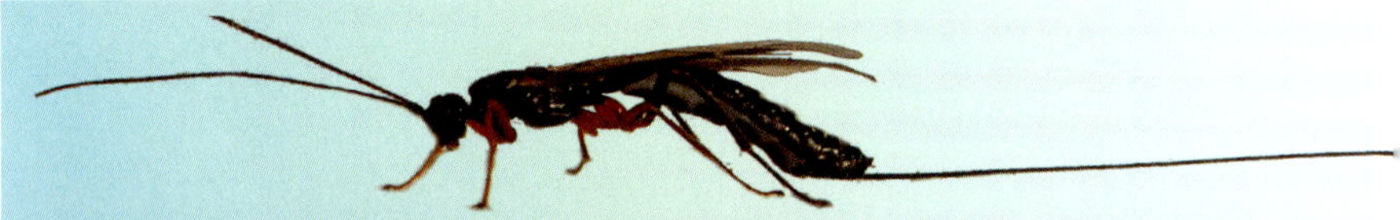

This ichneumonid female was observed searching for a host in coastal sand dunes. The host is possibly *Lacinipolia olivacea,* a noctuid larva tunneling in the sand. See p. 157.

Subfamily Campopleginae

The larvae of this subfamily are internal parasites of caterpillars and sawfly larvae.

Subfamily Diplazontinae

These Ichneumonids are internal parasites of aphid-eating flower fly larvae. The parasitoid egg is injected into the host egg or larva, and the adult emerge from the pupa.

Spider Wasp, Hawk Wasps
Family Pompilidae

Length: 3/8" to 5/8". These are black wasps with long legs, filiform antennae, dark wings and sometimes a contrasting color pattern on their body. These wasps feed their brood with paralyzed spiders stored in underground nests found where gravel or coarse sand is exposed.

Figitid Wasps
Figitdae Wasp
Subfamily Charipinae

Length: less than 1/32". These wasps are closely related to Cynipid wasps, but are parasitoids and hyperparasitoids on aphids.

***Charips* sp.**

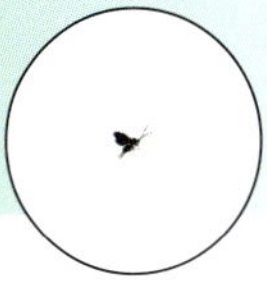

Gall Wasps
Family Cynipidae

Cynipid wasps induce galls on plants. Species of the closely related Charipidae family are hyperparasitoids on aphids.

Diplolepis polita

Length: 1/8". In Alaska, *Diplolepis polita* induce the formation of spiny spherical galls on rose leaves.

Cuckoo Wasps
Family Chrysididae

Omalus aenus

Length: 1/4". Adults are metallic blue, and the body surface is deeply sculpted. The larvae are external parasites of Penphredonidae wasps (see p. 168) *Penphredon* and *Passaloecus,* which either nest in pre-existing cavities of rotten wood or bore their own galleries in soft pithy stems of plants like elderberry, and stock up on aphids for their larvae.

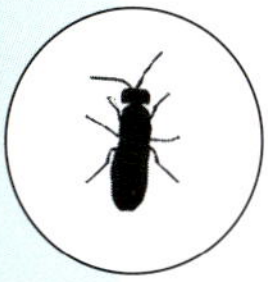

Sphecid Wasps. Family Sphecidae

Subfamily Pemphredoninae

These solitary wasps nest in the ground or in cavities left in the wood by beetle larvae. They bore tunnels in the soft pith of shrubs such as elderberry or in rotten wood. Most, if not all, are insect predators.

The larvae of this species are reared in a chamber in punky birch wood.

Crabro* sp. *(latipes or monticola)
Length: 3/8". The male's forelegs are expanded like a shield. The adults bore a tunnel in clay and stock up on killed prey, mostly diptera, to feed their larvae which develop in the underground chamber.

Entrance of *Crabro* sp. tunnel

Frontal view of the male's face and forelegs

Lateral view of the male's forelegs

Fully grown larva

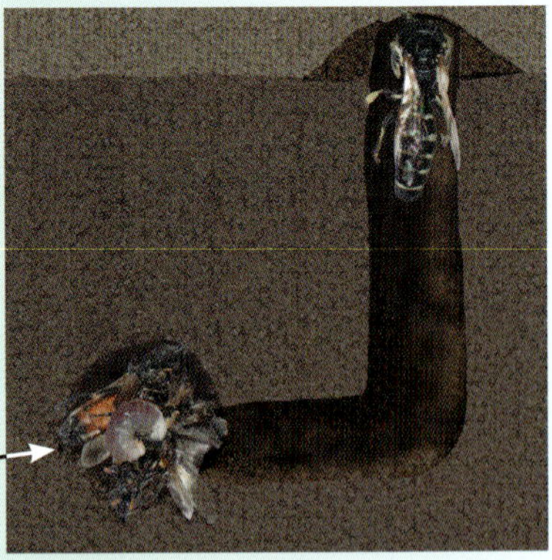

Cross-section of *Crabro* sp. tunnel with larvae surrounded by prey in the underground chamber.

Honey Bees, Bumblebees
Family Apidae

Most bees and bumblebees are social insects. They have "tongues" for licking the nectar out of the flowers instead of articulated mandibles like a wasp.

Female workers

2X

Drone (male)

Queen

Domestic Bee
Apis mellifera
Length: 1/2" to 5/8". This species survives only under domestication. New queens are usually imported each year to Alaska.

Bumblebee
***Bombus* sp.**
Length: 1/2" to 3/4". Queens overwinter and start a new colony early in the spring. In a sheltered space, they fabricate a few cells with wax they collected from flower glands. In these few cells, they first rear a batch of workers which are sterile females. As soon as these are adults, they take over the task of providing for the young. The queen's function is then reduced to egg laying. By the end of the summer, some of the emerging bumblebees are sexually developed and will overwinter.

Bumblebees are remarkably well adapted to cold weather and one species can even be found living in the high arctic.

2X

Paper Wasps, Yellowjackets
Family Vespidae

Vespula vulgaris
Length 1/2" to 3/4". These wasps are highly social and build paper-like nests in burrows in the ground or sheltered in a wood pile, for instance. Only the queens survive the winter. As soon as they emerge from hibernation, each queen builds a small paper nest and rears a first brood of workers. After these emerge, the workers take over the maintenance and expansion of the nest, and hunt to feed the next brood of larvae. The queen remains in the nest thereafter. Towards the end of the summer, the workers, now numbering several hundred, build larger cells in which males and queens are reared. Soon after, the queen dies and a new generation of queens and males emerge, disperse and mate. The males and the workers die off in the fall while the new queens find overwintering sites. These wasps hunt soft-bodied insects and scavenge for meat and ripe fruits. This last habit often brings them in conflict with humans.

3X

Queen

Vespula rufa
Length 5/8". This species' abdomen is black, pale white and orange-red. It lives only in the northern forest zone of North America. Nests have never been found.

2X

2X

Dolichovespula norvegicoides
Length: 3/4". This wasp builds a small paper nest 3-4 " in diameter above the ground, on the ground or in a cavity underground. This species is found in Alaska and in the northern half of Canada.

Male

1.5X

Dolichovespula maculata
Length: 3/4 ". The body of this species is black with a few pale yellow markings. They build paper nests perched off the ground that can measure more than a foot in diameter. This species is widespread throughout North America.

Nest of the aerial wasp
Dolichovespula arenaria

Potter Wasps
Family Eumenidae

These wasps are solitary. Some nest in decaying wood, in hollow twigs or fabricate clay nests attached to vegetation.

Symmorphus cristatus
Length: 1/2". This small wasp is a specialist predator of willow leaf beetles. The wasp makes a nest in abandoned beetle galleries in punky wood. The female lays one egg at the roof of a gallery, then fills that section of the gallery with dead willow leaf beetle larvae. It then seals that section of the gallery with clay and wood fibers. Then it starts a new chamber and repeats the same process. This species is found throughout North America.

***Zethus* sp.**

Ants
Family Formicidae

Twenty-four species and seven genera of ants are known to be in Alaska. They are widespread, and when numerous, influence the types of species of insects living in their immediate environment.

Formica sp.

Length: 1/4 ". These small ants nest in the ground. They provide for their larvae which are reared in underground chambers. During warm days, the workers bring the larvae close to the surface of the ground or under flat stones heated by the sun.

Worker

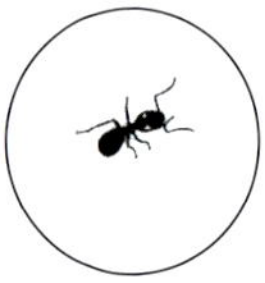

Carpenter Ant
Camponotus herculeanus

Length: Queens to 5/8". The large-winged queens and their smaller mates can be seen during their mating flight in mid summer. The workers are wingless.

In nature, carpenter ants act as recyclers of dead wood. The ants bore galleries in the decayed wood thereby accelerating the recycling of nutrients. For the same reason, they can be pests in structural lumber.

Male

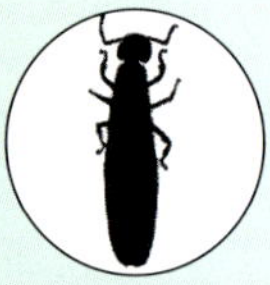

Queen

Carpenter ants overwinter as adults inside cavities bored in punky wood at the base of a tree such as this spruce tree.

Amphibious: living both in the water and in the air

Aphid mummy: inflated skin of an aphid after it has been parasitized (See p. 48)

Aposematic: warning colors (See p. 26)

Arthropod: invertebrates having articulated joints

Berlese funnel: device to collect live invertebrates from a soil sample (See p. 10)

Cambium: plant cell layer that gives rise to the phloem and heartwood (See p. 42)

Class: a division of the animal kingdom (See p. 78)

Compound eye: large eyes of insects composed of numerous ommatidia (See p. 22)

Cornicle: one of two tubes rising from the top of aphid abdomens.

Ectoparasite: external parasite (See p. 45)

Elytra: hardened front wings of beetles (See p. 19)

Endoparasite: internal parasite (See p. 45)

Entomophagous: feeding on insects.

Family: taxonomic division of class (See p. 78)

Feral: wild, or domesticated animals returned to the wild

Filliform: shaped like a wire, wiry

Fundatrix: wingless female aphid giving birth to live nymphs without prior mating (See p. 28)

Gall: abnormal growth on a plant tissue induced by a living organism (See p. 38)

Genus: group of closely related species (See p. 78)

Guild: group of insects with similar ecological functions (See p. 32)

Haltere: atrophied hindwing in Diptera (See p. 19)

Honeydew: sweet excretion of aphids or hoppers (See p. 57)

Host: animal on which a parasite feeds, plant food of a herbivore (See p. 45)

Hyperparasite, hyperparasitoid: parasite of a parasite (See p. 48)

Key: method used in biological science to identify an organism by process of elimination (See p. 79)

Lindgren trap: hanging funnels designed to capture tree-boring insects (See p. 11)

Malaise trap: screened tent-shaped insect trap (See p. 9)

Mandible: mouthpart

Meristem: actively dividing cells at the tip of a bud or stem

Metamorphosis: transformation of larva into adult form (See p. 21)

Mimicry: quality of resembling an organism unfit for consumption for gaining protection from predators (See p. 27)

Mutualistic: relationship that profits two or more organisms

Necrophagous: feeding on dead organisms

Niche: function of an organism in a community (See p. 32)

Nymph: immature stage of insects with incomplete metamorphosis (See p. 21)

Ocellus (plural: ocelli): simple eye (See p. 22)

Ommatidium: single light sensor of a compound eye (See p. 22)

Order: taxonomic division of a class (See p. 78)

Outbreak: unusual increase of an insect population

Parasite: species that feeds on one or several individual organisms without killing them (See p. 43)

Parasitoid: species that complete its development entirely on a single organism, usually killing it (See p. 43)

Pathogenic: causes disease

Petiole: narrow stem at the base of a wasp abdomen

Phenology: sequence and timing of development

Pheromone: chemical released by one species to communicate with other individuals of the same species (See p. 24)

Phloem: inner layer of the bark where capillaries carry the carbohydrates and other nutrients from the leaves or needles down to the growing areas of the plant

Polliny: sac containing pollen in orchids (See p. 41)

Predator: species that kills other species (See p. 43)

Proboscis: mouthpart shaped like an extended tube (See p. 20)

Pupa: resting transitional stage between larvae and adults in insects with complete metamorphosis (See p. 21)

Riparian: living along the shores of a lake or river

Saprophagous: feeding on decaying tissue

Serrated: toothed like a saw blade

sp.: acronym for "not identified to the species level"

Species: all the individuals that share the same gene pool (See p. 78)

spp.: abbreviated plural of species.

Alder
Alnus spp.
Insects: *Altica tombacina, Boemerina occidentalis, Campaea perlata, Chrysomela* sp., *Cladara limitaria, Clepsis persicana, Dicerca* sp., *Eriocampa ovata, Leucobrephos brephoides, Orgyia antiqua, Psylla floccosa, Sicya macularia, Trichiosoma triangulum, Trypodendron* sp.

Aspen
Populus tremuloides
Insects: *Campaea perlata, Deprana bilineata, Dorytomus* sp., *Phyllocnistis populiella, Phratora kenaiensis, Pterorous glaucus, Sicya macularia, Trypodendron* sp.

Beach pea
Lathyrus maritimus
Insects: *Lacinipolia olivacea*

Birch
Betula spp.
Insects: *Altica tombacina, Autographa* sp., *Campaea perlata, Cladara limitaria, Clepsis persicana, Deprana bilineata, Elasmostethus interstinctus, Energia infumata, Fenusa pusilla, Leucobrephos brephoides, Orgyia antiqua, Parasemia plantaginis, Parastichis suspecta, P. hudsonia, Polygonia faunus, Profenusa thomsoni, Rheumaptera subhastata, Sicya macularia, Trichiosoma triangulum, Trypodendron* sp., *Upis ceramboides.*

Bluegrass
Calamagrostis sp.
Insects: *Mayetiola sp.*

Calamus
Calamus sp.
Insects: *Plateumaris*

Clover
Trifolium pratense
Insects: *Colias philodice*

Cotoneaster
Cotoneaster sp.
Insects: *Orgyia antiqua*

Cottonwood, balsam poplar
Populus trichocarpa, P. balsamifera
Insects: *Deprana bilineata, Dorytomus* spp., *Energia infumata, Furcula occidentalis, Leucobrephos brephoides, Nymphalis antiopa, Pemphigus* spp., *Pterorous glaucus, Scoliopterix libatrix, Trichiosoma triangulum, Trypodendron* sp.

Cow parsnip
Heracleum lanatum
Insects: *Judolia montivagans, Tenthredo* sp., *Xestoleptura behrensi*

Currant
Ribes spp.
Insects: *Sicya macularia*

Dandelion
Taraxacum
Insects: *Phragmatobia fuliginosa*

Delphinium
Delphinium sp.
Insects: *Polychrysia esmeralda*

Dogwood
Cornus stolonifera
Insects: *Altica tombacina*

Fireweed
Epilobium angustifolium, E. latifolium
Insects: *Albuna pyramidalis, Altica tombacina, Alypia langtoni, Hyles galii, Bromius obscurus, Mompha* sp.

Highbush cranberry
Viburnum edule
Insects: *Hemaris thysbe*

Horsetail
Equisetum spp.
Insects: *Grypus equiseti, Hippuriphila mancula,*

Kinnikinnik
Arctostaphyllos uva-ursi
Insects: *Tamalia coweni*

Labrador tea
Ledum palustre
Insects: *Celestrina lucia?*

Mountain ash
Sorbus
Insects: *Archips rosaceana*

Larch
Larix spp.
Insects: *Pristiphora ericksonii,*

Nettle
Urtica spp.
Insects: *Aglais milberti*

Prickly rose
Rosa acicularis
Insects: *Diplolepis polita, Altica tombacina*

Plantain
Plantago sp.
Insects: *Parasemia plantaginis*

Plum-cherry
Prunus sp.
Insects: *Archips conflictana*

Rush
Juncus sp.
Insects: *Plateumaris* sp.

Saxirage
Saxifraga spp.
Insects: *Parnassius phoebus*

Sedge
Carex spp.
Insects: *Plateumaris* sp.

Strawberry
Fragaria sp.
Insects: *Altica tombacina, Bromius obscurus, Clepsis trileucana*

Spruce
Picea spp.
Insects: *Adelges spp., Arhopalus productus, Choristoneura conflictana, C. fumiferana, Clepsis persicana, Dendroctonus rufipennis, Dicerca* sp., *Elatobium abietinum,. Ips* spp., *Monochamus scutellatum, Melanophila fulvoguttata, Oxypteris acuminata, Phymatodes dimidiatus, Pineus* spp., *Pogonocherus penicillatus, Polygraphus rufipennis, Pronocera collaris, Sirex cyaneus, Tetropium parvulum, Trichiosoma triangulum, Trypodendron* sp., *Xylotrechus undulatus.*

Blueberry
Vaccinium spp.
Insects: *Orgyia antiqua, Parasemia plantaginis, Sicya macularia.*

Willow
Salix spp.
Insects: *Altica tombacina, Bromius obscurus, Cacopsylla alaskensis, Campaea perlata, Chrysomela* spp., *Cladara limitada, Dasineura* spp., *Dorytomus* spp., *Energia infumata, Euura* spp., *Furcula occidentalis, Iteomyia* sp., *Lepyrus gemellus, Leucobrephos brephoides, Nymphalis antiopa, Orgyia antiqua, Parastichis suspecta, Phratora interstitialis, Phyllocolpa* spp. *Polygonia faunus, Pontania* spp., *Pterocomma sp., Rabdophaga rosaria, R. rigidae, R. salicis, Saperda* sp., *Scoliopterix libatrix, Sicya macularia, Trichalophus alternatus.*

Wormwood
Artemisia spp.
Insects: *Papilio machaon*

For more information about insects

In this Internet age, it is possible to access information like never before, even from the remotest village in Alaska. When I identified an insect in this guide, I used an Internet browser and typed as key words the family, genus and/or species and searched the web. For instance I found a yellowjacket wasp. I saw that in this guide it belonged to the family Vespidae, and I think it might be *Dolichovespula norvegicoides*. I type "Vespidae key Dolichovespula norvegicoides" and get 3 hits including 2 illustrated keys to the genera and species of yellowjacket wasps.

If I want a picture, I search "images" by entering the genus and species (3 hits) or only the genus (478 hits). Now, yellowjackets must be a popular subject, but it is worth the try for any insect. There is no guarantee that insects on the images posted on the web are correctly identified, but there is a lot of good material out there, hosted by folks who are dedicated to creating a virtual insect collection. The web is also a place to gather information about the known biology or the distribution of a species.

Additional information on insects of south-central Alaska can be obtained from:

- Forest Service State and Private Forestry (located in Anchorage) 907-743-9455 FAX: 907-743-9479. 3301 C Street, Suite 202, Anchorage, AK 99503-3956

- University of Alaska Fairbanks Cooperative Extension Service which has offices in south-central Alaska in Anchorage, Kodiak, Palmer, Sitka, Palmer, Kenai/Soldotna

- The Alaska Entomological Society encourages any aspiring entomologist to join as member. Details can be found at http://www.akentsoc.org

- To help document the Alaskan insect fauna, the University of Alaska Museum in Fairbanks actively requests donations of insect collections from anywhere in the state, especially from off the road system. If you have the opportunity to collect insects at remote sites, please contact the Insect Collection Curator at the Museum of the North in Fairbanks at ffdss@uaf.edu.

Useful Books

The main purpose in writing this guide is to provide a source of information on insects of south-central Alaska. Few general insects guides are available for this area. Înformation in this guide has been compiled from various sources, including private and state collections, and collection and specialised literature.

The following books are useful:

Borror, D. J. and D. M. DeLong. 1954
An Introduction to the Study of Insects. Holt, Rhinehart and Winston.

Borror, D. J. and R. E. White. 1992.
A Field Guide to the Insects of North America.
Peterson Field Guide Serie. Houghton Mifflin Company, Boston.

Holsten, E., Hennon, P., L. Trummer, and M. Schultz. 2001.
Insects and Diseases of Alaskan Forests. USDA, Forest Service Alaska Region.

Hudson, J. and R. H. Armstrong. 2005.
Dragonflies of Alaska. Todd Communication.

Furniss, R. L. and Carolin, V. M. 1977.
Western Forest Insects. Misc. publication US Dept Agric. Forest Service.

Solomon, J. D., 1995.
Guide to the Insect Borer of North American Broadleafs and Shrubs.
Agric. handbook, USDA

Tilden, J. W. and A. C. Smith. 1986.
A Field Guide to Western Butterflies.
Peterson Field Guide Series. Houghton Mifflin Company, Boston.

White, R. E. 1983.
A Field Guide to the Beetles of North America.
Peterson Field Guide Series. Houghton Mifflin Company, Boston.

Merrit R.W. and K. W. Cummins, Eds. 1987.
An Introduction to the Aquatic Insects of North America.

Agriculture Canada has published a number of books on the insects of Canada and Alaska.

Notes

Notes

Index

C

D

E

F

G

H

N

O

P

R

S

T

U

V

W

X

Y

Z